# Frontiers in Polar Biology in the Genomic Era

WITHDRAWN

Committee on Frontiers in Polar Biology
Polar Research Board
NATIONAL RESEARCH COUNCIL
*OF THE NATIONAL ACADEMIES*

THE NATIONAL ACADEMIES PRESS
Washington, D.C.
**www.nap.edu**

THE NATIONAL ACADEMIES PRESS   500 Fifth Street, NW   Washington, DC 20001

NOTICE: The project that is the subject of this report was approved by the Governing Board of the National Research Council, whose members are drawn from the councils of the National Academy of Sciences, the National Academy of Engineering, and the Institute of Medicine. The members of the committee responsible for the report were chosen for their special competences and with regard for appropriate balance.

This material is based upon work supported by the National Science Foundation under NSF Grant No. OPP-0132773, Master Agreement No. 9911018. Any opinions, findings, conclusions, or recommendations expressed in this publication are those of the author(s) and do not necessarily reflect the views of the National Science Foundation.

International Standard Book Number 0-309-08727-9 (Book)
International Standard Book Number 0-309-51229-8 (PDF)
Library of Congress Catalog Card Number 2003093146

Cover design by Van Nguyen.

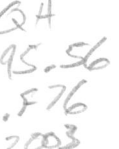

Additional copies of this report are available from the National Academies Press, 500 Fifth Street, N.W., Lockbox 285, Washington, DC 20055; (800) 624-6242 or (202) 334-3313 (in the Washington metropolitan area); Internet, http://www.nap.edu

Copyright 2003 by the National Academy of Sciences. All rights reserved.

Printed in the United States of America.

# THE NATIONAL ACADEMIES
*Advisers to the Nation on Science, Engineering, and Medicine*

The **National Academy of Sciences** is a private, nonprofit, self-perpetuating society of distinguished scholars engaged in scientific and engineering research, dedicated to the furtherance of science and technology and to their use for the general welfare. Upon the authority of the charter granted to it by the Congress in 1863, the Academy has a mandate that requires it to advise the federal government on scientific and technical matters. Dr. Bruce M. Alberts is president of the National Academy of Sciences.

The **National Academy of Engineering** was established in 1964, under the charter of the National Academy of Sciences, as a parallel organization of outstanding engineers. It is autonomous in its administration and in the selection of its members, sharing with the National Academy of Sciences the responsibility for advising the federal government. The National Academy of Engineering also sponsors engineering programs aimed at meeting national needs, encourages education and research, and recognizes the superior achievements of engineers. Dr. Wm. A. Wulf is president of the National Academy of Engineering.

The **Institute of Medicine** was established in 1970 by the National Academy of Sciences to secure the services of eminent members of appropriate professions in the examination of policy matters pertaining to the health of the public. The Institute acts under the responsibility given to the National Academy of Sciences by its congressional charter to be an adviser to the federal government and, upon its own initiative, to identify issues of medical care, research, and education. Dr. Harvey V. Fineberg is president of the Institute of Medicine.

The **National Research Council** was organized by the National Academy of Sciences in 1916 to associate the broad community of science and technology with the Academy's purposes of furthering knowledge and advising the federal government. Functioning in accordance with general policies determined by the Academy, the Council has become the principal operating agency of both the National Academy of Sciences and the National Academy of Engineering in providing services to the government, the public, and the scientific and engineering communities. The Council is administered jointly by both Academies and the Institute of Medicine. Dr. Bruce M. Alberts and Dr. Wm. A. Wulf are chair and vice chair, respectively, of the National Research Council.

**www.national-academies.org**

## COMMITTEE ON FRONTIERS IN POLAR BIOLOGY

**H. WILLIAM DETRICH III** (*Chair*), Department of Biology, Northeastern University, Boston, Massachusetts
**JODY W. DEMING**, School of Oceanography, University of Washington, Seattle
**CLAIRE FRASER**, The Institute for Genomic Research, Rockville, Maryland
**JAMES T. HOLLIBAUGH**, Department of Marine Sciences, University of Georgia, Athens
**WILLIAM W. MOHN**, Department of Microbiology and Immunology, University of British Columbia, Vancouver
**JOHN C. PRISCU**, Department of Land Resources and Environmental Sciences, Montana State University, Bozeman
**GEORGE N. SOMERO**, Hopkins Marine Station, Stanford University, Pacific Grove, California
**MICHAEL F. THOMASHOW**, Michigan State University, East Lansing
**DIANA H. WALL**, Natural Resource Ecology Laboratory, Colorado State University, Fort Collins

*Staff*

Evonne Tang, Study Director
Ann Carlisle, Administrative Associate
Bryan Ericksen, PRB Intern (January 14, 2002 through May 3, 2002)

## POLAR RESEARCH BOARD

**ROBIN BELL** (*Chair*), Lamont-Doherty Earth Observatory, Palisades, New York
**MARY ALBERT**, Cold Regions Research and Engineering Laboratory, Hanover, New Hampshire
**RICHARD B. ALLEY,** Pennsylvania State University, University Park
**AKHIL DATTA-GUPTA**, Texas A&M University, College Station
**GEORGE DENTON**, University of Maine, Orono
**HENRY P. HUNTINGTON**, Huntington Consulting, Eagle River, Alaska
**DAVID W. KARL**, University of Hawaii, Honolulu
**MAHLON C. KENNICUTT**, Texas A&M University, College Station (ex-officio)
**AMANDA LYNCH**, University of Colorado, Boulder
**W. BERRY LYONS**, Byrd Polar Research Laboratory, Columbus, Ohio
**ROBIE MACDONALD**, Fisheries and Oceans Canada, Institute of Ocean Sciences, British Columbia
**MILES MCPHEE**, McPhee Research Company, Naches, Washington
**CAROLE L. SEYFRIT**, Old Dominion University, Norfolk, Virginia
**PATRICK WEBBER**, Michigan State University, East Lansing (ex-officio)

*Staff*

Chris Elfring, Director
Sheldon Drobot, Staff Officer
Evonne Tang, Staff Officer (on loan from the Board on Life Sciences)
Ann Carlisle, Administrative Associate

# Preface

The polar biological sciences stand on the threshold of a revolution, because the availability of genome sciences and other new enabling technologies has opened the door to study an incredible array of important questions and issues, both fundamental and practical. As this revolution occurs, we will enter a new era that holds the promise of a greatly enhanced understanding of polar ecosystems, their biodiversity, and the interactions of their constituent organisms and communities. Genome sciences is "the study of the structure, content, and evolution of genomes," including the "analysis of the expression and function of both gene and proteins." These approaches will help us to determine not only the organisms that are present in polar biomes but also to look into their evolution to thrive in extreme cold, their interactions as biological systems, and their capacity to handle global change. Our knowledge of polar ecosystems will become not only global in scope but also more mechanistic in explanation.

The Committee on Frontiers in Polar Biology was charged to examine the opportunities and challenges for conducting research on Arctic and Antarctic organisms and ecosystems using genomic technologies. This study was requested by the Office of Polar Programs and the Biology Directorate of the National Science Foundation. Encompassing both poles and essentially all biological disciplines, the task was both stimulating and difficult. I am deeply indebted to my colleagues on the committee, whose breadth of expertise, insights, and selfless efforts have driven this study idea to reality. On behalf of the committee, I also wish to extend my gratitude to the many biologists (see Appendix C) of all disciplines who participated in our workshop on Frontiers in Polar Biology (September 9-11, 2002). Their ideas and thoughtful criticism contributed significantly to shaping the committee's thinking and to the generation of this report.

This report opens with an Introduction (Chapter 1) to genomics, the paleogeological and paleoclimatic history of the polar regions, the unique features that shape biological and evolutionary processes in polar regions, and a summary of the applicability of genomic technologies to polar biology. Subsequently, we define the key questions in polar biology that are amenable to genomic approaches (Chapter 2); identify critical species whose genomes, transcriptomes, proteomes, and metabolomes should be analyzed under the aegis of a Polar Genome Science Initiative (Chapter 3); develop strategies to facilitate interactions and technology transfer within the polar biological community, between polar and nonpolar biologists, and through educational outreach to academic and lay audiences (Chapter 4); and discuss other technologies, facilities, and infrastructure whose development would complement the genome initiative (Chapter 5). The report concludes with the committee's findings and recommendations for implementation of these ideas (Chapter 6).

The report's recommendations, if implemented, will require increased funding by the National Science Foundation; and we recognize that this may not be an easy decision. Nevertheless, the committee presents an ambitious research agenda. We believe that we are ready to move forward: Prior genome projects have developed a strong infrastructure and considerable talent, and the costs for genome-related science are rapidly dropping. We believe that in the near future the commonly heard question "What is the justification for sequencing the genome of organism *Genome species*?" will shift to the imperative, "You must sequence the genome of your organism(s) (or metagenome of an environmental community) so that you obtain a comprehensive understanding of your biological system." The competitive peer review system of the National Science Foundation will ensure that the most rigorous and relevant genome projects are supported and that results of the highest quality are disseminated widely.

This report would not exist but for the heroic efforts of the supporting staff of the Polar Research Board: Evonne Tang, Ann Carlisle, Bryan Erickson, and Chris Elfring. They ensured that the committee stayed focused "on goal" through their expertise, tireless support, and efficient marshaling of resources. I believe I speak for the committee when I say that their assistance guaranteed that our participation in this study was both successful and delightful.

H. William Detrich III, Chair<br>
Committee on Frontiers in Polar Biology

# Acknowledgments

This report is a product of the cooperation and contribution of many individuals. The committee would like to thank all the participants of the Frontiers in Polar Biology Workshop on September 9 and 10, 2002 (see Appendix C) and the following individuals for their input:

Don K. Button, University of Alaska, Fairbanks
Glenn F. Cota, Old Dominion University
Kenneth Dunton, University of Texas at Austin
Wade H. Jeffrey, University of West Florida
John Lisle, U.S. Geological Survey
Harlan Miller, University of Georgia
Jill A. Peloquin, College of William and Mary
Scott Rogers, Bowling Green State University
Walker O. Smith, College of William and Mary
Warwick F. Vincent, Université Laval
Patricia Yager, University of Georgia

This report has been reviewed in draft form by persons chosen for their diverse perspectives and technical expertise, in accordance with procedures approved by the National Research Council's Report Review Committee. The purpose of this independent review is to provide candid and critical comments that will assist the institution in making its published report as sound as possible and to ensure that the report meets institutional standards for objectivity, evidence, and responsiveness to the study charge. The review comments and draft manuscript remain confidential to protect the integrity of the deliberative process. We wish to thank the following individuals for their review of this report:

Brian Barnes, University of Alaska, Fairbanks
Andrew Cossins, The University of Liverpool, United Kingdom
David M. Karl, University of Hawaii, Manoa
Norman Huner, University of Western Ontario, Canada
Robie Macdonald, Fisheries and Oceans Canada, British Columbia
Adam G. Marsh, University of Delaware, Lewes
Norman Pace, University of Colorado, Boulder
Lyle G. Whyte, McGill University, Quebec, Canada

Although the reviewers listed above have provided many constructive comments and suggestions, they were not asked to endorse the conclusions or recommendations, nor did they see the final draft of the report before its release. The review of this report was overseen by Christopher R. Somerville, Director of the Department of Plant Biology, Carnegie Institute of Washington. Appointed by the National Research Council, he was responsible for making certain that an independent examination of this report was carried out in accordance with institutional procedures and that all review comments were carefully considered. Responsibility for the final content of this report rests entirely with the authoring committee and the institution.

# Contents

| | | |
|---|---|---|
| | SUMMARY | 1 |
| 1 | INTRODUCTION | 15 |
| | What Is Genomics? 15 | |
| | Geologic and Climatic Trends That Influenced Evolution in the Polar Regions, 16 | |
| | Physical Parameters That Shape Biological Processes, 21 | |
| | Evolution in Polar Regions, 22 | |
| | Examples of Research Areas That Could Be Addressed with Genomic Tools, 22 | |
| 2 | IMPORTANT QUESTIONS IN POLAR BIOLOGY | 25 |
| | Evolution and Biodiversity of Polar Organisms, 26 | |
| | Polar Physiology and Biochemistry, 53 | |
| | Polar Microbial Communities, 64 | |
| | Human Impacts, 73 | |
| | Summary, 80 | |
| 3 | THE POLAR GENOME SCIENCE INITIATIVE | 82 |
| | Selection of Organisms and Consortia for Genome Analysis, 82 | |
| | Structure of a Polar Genome Project, 93 | |
| | Creation of a Polar Genome Science Initiative, 102 | |

| 4 | COMPLEMENTS TO GENOME SCIENCE: ENABLING TECHNOLOGIES, FACILITIES, AND INFRASTRUCTURE<br>Enabling Technologies, 105<br>Facilities and Infrastructure, 109 | 105 |
|---|---|---|
| 5 | AN INTEGRATED POLAR BIOLOGY COMMUNITY: INTERACTIONS AMONG SCIENTISTS, EDUCATION, AND OUTREACH<br>Facilitating Interactions and Technology Transfer Across Scientific Disciplines, 119<br>Education and Outreach, 123 | 119 |
| 6 | FINDINGS AND RECOMMENDATIONS<br>A New Unifying Approach to Polar Biological Research, 128<br>Coordination Is Essential, 129<br>Virtual Genome Science Centers, 130<br>Enabling Technologies, 130<br>Increasing Awareness and Education, 130<br>Impediments to Integrated Polar Science, 131 | 128 |

REFERENCES 133

APPENDIXES
A  Committee Member Biosketches 157
B  Workshop on Frontiers in Polar Biology—Agenda 161
C  Workshop on Frontiers in Polar Biology—Participants 163
D  Acronyms 164

# Summary

As we enter the twenty-first century, the polar biological sciences stand well poised to address numerous important issues, many of which were unrecognized as little as 10 years ago. From the effects of global warming and elevated ultraviolet radiation on polar organisms to the potential for life in subglacial Lake Vostok, the opportunities to advance our understanding of polar ecosystems are unprecedented. At the same time, the biological sciences are in the midst of a major change in technological capabilities. The era of "genome-enabled" biology is upon us, and these new technologies will allow us to examine polar biological questions of unprecedented scope and to do so with extraordinary depth and precision.

All polar biological disciplines with applicability to polar regions, including systematics, microbiology, ecology, evolutionary biology, physiology, biochemistry, and molecular biology, will be transformed by exploiting the new technologies available to biologists. These genome-enabled methods will allow us to examine the genomic structure of organisms and communities, monitor changes in the expression of genes, and obtain detailed images of how the physiologies of organisms are affected by natural or anthropogenic changes in the environment. Complementing the broad array of technologies associated with genomics are other new enabling technologies, such as platforms equipped to monitor biomes in real time and autonomous underwater vehicles that will expand our spatial and temporal understanding of marine and terrestrial communities and their dynamics. The new approaches and tools are not ends in themselves but pathways into a new frontier of science opportunities.

The Committee on Frontiers in Polar Biology (see Appendix A for committee membership) examined the opportunities and challenges of using the new technologies and methods of biology to conduct research on key questions related to Arctic and Antarctic organisms. Specifically, the study committee was given the following charge:

- Identify high-priority research questions that can benefit most from the new tools of biology in polar regions and recommend ways to facilitate and accelerate the transfer and use of genomic technologies to answer fundamental questions about Arctic and Antarctic organisms.
- Discuss the potential applications of genomic sciences and functional genomics to molecular biology, microbiology, biochemistry, physiology, evolutionary processes, and microbial ecology in polar regions and identify the need for development of new technologies or methods specifically for polar regions.
- Seek ways to facilitate increased interaction between biological scientists working in polar regions and other biological scientists.
- Assess impediments to the conduct of polar genomic research, such as issues related to facilities, infrastructure, and maintenance of biological sample collections and issues related to manpower and education needs.

The committee conducted its analysis in a series of meetings. To gather input from the wider scientific community, the committee organized a special two-day Workshop on Frontiers in Polar Biology (see Appendixes B and C for workshop agenda and list of participants). Workshop participants included biologists with expertise in microbial, protistan, soil, plant, invertebrate, fish, bird, and mammalian systems; in paleobiology and astrobiology; in genomics and bioinformatics; and in education and outreach. To exploit synergies of perspective, the workshop was also balanced among biologists who conduct research in the Arctic, in the Antarctic, and in nonpolar environments. The breadth of expertise helped the committee identify key science questions and assess the opportunities and challenges associated with exploiting new genomic (and complementary nongenomic) technologies in polar biological research.

## WHY POLAR BIOLOGY?

Clearly, biology in polar regions shares its traditions with all of biology—it explores the same fundamental questions about organisms and ecosystems, from essentially the same diversity of disciplinary perspectives and using similar methods. Like the rest of biology, polar biology has changed over time, because of both the increasing sophistication of our knowledge base and the advances in technology and computing power. But different regions of Earth have always offered their own compelling scientific opportunities and scientists often specialize so that they can explore these opportunities in depth.

For the polar regions, great distances, physical isolation, long periods of darkness, and extreme climates have always posed special challenges—the challenges associated with getting to and operating in these environments and of gaining a full perspective on a species when observations were limited to the "good" months when it is light and warm enough to conduct work. Thus, improvements in technology have always tied closely with advances in our ability to conduct science in polar regions and to expand the range of questions that could be addressed. Although some of the key research questions identified in this report indeed apply to other regions, there are reasons for the special polar emphasis. First, in general, polar regions remain one of the least studied and least understood ecosystems on the planet, and improving this understanding becomes more and more critical as we come to see the relationship of the polar regions with global processes. Second, and more specifically, genome research applied to polar biology would serve as a useful "test bed" for temperate and tropical regions (e.g., there are tens of thousands of tropical fishes but only about 250 in Antarctica so our ability to develop a comprehensive view is enhanced). Finally, in the end we need thorough understanding of all ecosystems so that we can conduct comparative studies across latitudinal clines, and these studies can then help elucidate physiological and biochemical mechanisms for adaptation in a way that no single perspective would allow.

Polar ecosystems provide study systems that can yield major insights into a wide range of basic and applied issues in biological science. The distinct geological, oceanographic, and climatic histories of the Arctic and Antarctic have created two polar ecosystems that differ in some attributes while sharing similarities in others. The Arctic, in essence, is an ice-dominated ocean surrounded by large, continental landmasses with wide, shallow continental shelves; the Antarctic, on the other hand, is a glaciated continent featuring narrow, deep coastal margins and surrounded by an ice-covered ocean. Both polar ecosystems are predominantly cold, isolated, and subject to pronounced seasonal cycles of temperature and photoperiod. Organisms not only survive, but thrive, under these extreme conditions, thus providing a unique perspective on the fundamental characteristics of life processes and the mechanisms of evolutionary adaptation. Many of the potential discoveries to be made in the study of adaptations of polar organisms stand not only to make important contributions to basic biological science but also to offer opportunities for advancing biotechnology and biomedicine—for instance in the development of protocols for cryopreservation of cells, tissues, and other biological materials.

Although remarkably well adapted to extreme physical conditions, polar organisms are highly sensitive to anthropogenic perturbation, such as the production of greenhouse gases and ozone-destroying chemicals. Human activities are already affecting polar ecosystems dramatically, and these effects are likely to increase in the future. If we are to predict the impact of environmental change on global ecosystems, it is critical that we evaluate closely and understand polar ecosystems, which in many ways may serve as "canaries in the coal mine" in terms of providing warnings about the effects of climate change worldwide.

## IMPORTANT QUESTIONS IN POLAR BIOLOGY

The committee identified four major focal areas of polar biological research that could benefit significantly from the application of genome-enabled technologies. This categorization reflects the important contributions that new technologies can make at all levels of biological organization, ranging from fundamental molecular-scale phenomena at the level of the genome to processes involving entire ecosystems and the human communities that depend on them.

1. *Evolution and biodiversity of polar organisms.* A major shaping force in the evolution of polar organisms and polar ecosystems was the development of the extreme physical conditions of the two polar regions, notably their very low temperatures. Polar species thus provide exceptional models for analyzing adaptive evolutionary change in extreme environments. Because the Arctic and Antarctic regions underwent glaciation during different geological epochs (Pleistocene and Miocene, respectively), comparison of adaptations in ecotypically equivalent boreal and austral taxa will provide important insights into convergent and divergent evolutionary adaptation. Permafrost, subglacial lakes, and other frozen environments may preserve a repository of ancient organisms and DNA that could be used (1) for analyses of biodiversity during different geological time periods and (2) to elucidate the evolutionary relationships between ancient and present-day organisms.

- What new types of genetic information have been gained to enable polar organisms to function well under the stressful physical conditions of the polar regions, especially extremes of cold? What "raw material" was used to fabricate the new types of genes found in cold-adapted polar organisms? Have Arctic and Antarctic species exploited similar or different genetic "raw material" to fabricate adaptations?

- What types of genetic information have been lost during evolution under extremely stable thermal conditions, such as those found in major regions of the polar seas? Are some polar organisms especially susceptible to global warming as a result of having lost genetic information needed to allow adaptation to higher and more variable temperatures? Do polar species contain the genetic information needed to cope with anthropogenic stresses such as ozone depletion?
- How rapidly do the genomes of polar organisms evolve and what are the mechanisms of genomic change?
- How do the processes of gene transcription and protein translation compare in polar and nonpolar species? Do polar species manifest reduced capacities to alter gene expression in the face of environmental change?
- What is the evolutionary origin of the organisms present in the polar ice caps, glaciers, and subglacial lakes? Are they metabolically active; and if so, do they possess novel metabolic pathways?
- Are polar environments reservoirs of paleogenes that can facilitate the evolution of present-day species through lateral gene transfer?
- What are the determinants of microbial diversity in the marine and terrestrial ecosystems of the polar regions?

2. *Polar physiology and biochemistry.* The abilities of polar organisms to carry out the physiological and biochemical processes required for metabolism, growth, and reproduction under extreme climatic conditions are based on widespread adaptive change. Proteins, membranes, and other key biochemical components of polar species exhibit a broad suite of adaptations that may at once "fit" their biochemistry to polar conditions and at the same time limit their functional range to the extreme environments of the poles. Among the important questions related to physiological and biochemical systems are the following:

- How have many "cold-blooded" polar fishes and invertebrates succeeded in reducing their metabolic rates? Can these mechanisms be used in biotechnological and biomedical procedures?
- What evolutionary mechanisms are available to adapt/preserve enzymatic activity at low temperatures?
- Can insights from study of these cold-adapted proteins guide biotechnological development of commercially useful molecules; for example, enzymes with novel activities that work efficiently at low and moderate temperatures?
- What are the types of molecules that serve as "antifreeze" agents and ice-nucleators? How do these molecules work? What biotechnological and biomedical potential is represented by these molecules?

- What types of physiological and biochemical adaptations enable the cells and organs of some small Arctic mammals to survive at sub-zero temperatures during hibernation? Can these same mechanisms be exploited in biomedical procedures, including cryosurgery and cryostorage of tissues and organs?
- What sensing and regulatory pathways have polar organisms evolved to cope with abiotic stresses?

3. *Polar microbial communities.* A broad suite of questions can be addressed to advance our understanding of the functioning of polar microbial ecosystems. To these ends, appropriate genomic methodologies must be linked to a variety of new methods for field investigation, including new drilling technologies and unmanned observatories. The potential rewards of these new lines of study are vast and include contributions to aquatic, terrestrial, and potentially, extraterrestrial biology. For example, the cryptoendolithic organisms and those dwelling in the perennially ice-covered lakes of the McMurdo Dry Valleys (which were once thought to be abiotic) have long been recognized as potential analogues of life (if any) on Mars. Similarly, the long-isolated microbial communities of Lake Vostok might serve as a model for evaluating the potential for life on Europa. Genome analysis of these organisms will provide us with an understanding of their origins and of genetic traits that might be expected in extraterrestrial life. Some key questions on polar ecosystem biology that can be addressed by creative use of genomics and other new technologies are the following:

- What types of microorganisms are present in polar aquatic and terrestrial ecosystems and what roles do these microorganisms play in ecosystem processes?
- What is the relationship between the composition and biogeochemical function of polar microbial communities?
- What factors control the composition, interaction, and productivity of the organisms in polar microbial aggregates?
- What is the lower temperature limit for evolving microbial life?
- Can we exploit an understanding of microbial life in Earth's polar regions to design probes and experiments to detect potential life in extraterrestrial environments?

4. *Assessment and remediation of human impact on polar ecosystems.* The impacts of human activities on polar ecosystems range from the direct impacts of activities such as fishing to less direct effects due to atmospheric modifications (greenhouse gases and ozone-destroying chemicals). The use of genomic technologies and other new approaches will

SUMMARY                                                                  7

yield important insights into all spatial and temporal scales of human effects and may provide strategies for the remediation of human perturbation of polar environments. Among the questions that could be addressed through application of these new technologies are the following:

- How will the thinning and shrinking of ice cover in Arctic marine habitats affect ecosystem structure?
- How will global warming influence the distributions of marine animal and plant species, for example, in the case of Arctic fishes that may have to remain in cold waters to find adequate nutrition?
- How will the loss of key species in polar communities affect ecosystem functioning (e.g., net primary productivity, decomposition rates, carbon and nitrogen cycles) and community composition at temporal and spatial (local to landscape) scales?
- Will climate change increase the frequency and success of biological invasions? What ecosystem-wide changes will be caused by these invasions?
- What is the potential of using genetic technologies as forensic tools? The utility of DNA methods in "environmental forensics" has been proven, for example, in the case of identifying species of whale meat sold commercially. Can these technologies be used for improved management of polar resources?
- What roles do soil microorganisms play in bioremediation in polar regions—for instance, in soils contaminated by petroleum spills? Can new varieties of microorganisms be obtained from contaminated soils for use in bioremediation processes?

To tap the full potential of genomics in addressing these and other key science questions, focused effort will be needed. The establishment of a Polar Genome Science Initiative would help address these research questions effectively. The Polar Genome Science Initiative would need to include the following aims:

- generation of genetic and physical maps of genomes of selected polar species;
- high-throughput sequencing of genomic DNA and expressed genes;
- gene identification and annotation;
- population analysis via single-nucleotide polymorphisms; and
- transcriptome, proteome, metabolome, and envirogenome analyses.

Because genome projects produce a high volume of sequence data and annotated information, a comprehensive Polar Genome Science Initiative must make provision for creation, curation, validation, and management of these databases and for bioinformatics tools necessary for insightful genome analyses.

Selecting polar organisms for genome analysis needs careful thought to ensure that resources are targeted effectively. Choices should be based on evidence that:

- analysis of its genome will address broad and significant scientific questions;
- it is a good model for evolution in an isolated polar environment;
- it provides opportunities for comparisons to organisms of comparable ecotype from polar habitats and along polar-to-temperate latitudinal clines; and/or
- its cellular processes possess characteristics of biotechnological interest.

Monitoring physiological and biochemical processes of polar organisms and monitoring polar ecosystems are keys to linking data generated from a Polar Genome Science Initiative to understanding and predicting organismal and ecosystem responses to environmental changes. Examples of novel approaches and advanced technologies for measuring biological processes include:

- multiple-element and compound-specific stable isotope analyses for studying photosynthetic and biogeochemical processes, respectively;
- stable isotope probing, an advanced culture-independent technique for isolation of DNA from microorganisms at a species level;
- instrument packages and "tags" for measuring geoposition, water depth, heart rate, and blood chemistry of animals that are subsequently released back in the field; and
- "biosensors" to detect particular DNA molecules or antigens for characterizing the compositions of aquatic microbial communities or for tracking plankton blooms.

## COMPLEMENTS TO GENOMIC SCIENCE: ENABLING TECHNOLOGIES, FACILITIES, AND INFRASTRUCTURE

The success of future polar genomic research depends not only on the new technologies available and the expertise of individual researchers but also on the equipment, infrastructure, and facilities that will enable researchers to sample, analyze, and experiment with organisms in polar ecosystems. The committee identified key technologies, infrastructure, and facilities that have to be developed or improved to facilitate the advancement of polar genomic research.

*Sampling.* Subglacial lakes that have been isolated from direct gas exchange with the atmosphere for perhaps 20 million years offer an incredible research opportunity. Clean technology must be developed to avoid contaminating the lakes with contemporary microbiota. Sampling procedure can also be improved by the development of new fast-access drilling methods and of ice-traverse technology to enable efficient field operations. For marine research, improved technologies for collection and shipment of sensitive specimens must be developed. Notably, the ability to reliably preserve and ship sensitive samples to be used in molecular biological and chemical analyses to home laboratories is essential to many research programs.

*Facilities.* To facilitate genomic research in the Arctic, improved facilities for collection, analysis, and shipment of materials are needed. The Toolik and Barrow facilities of Alaska are operating at, or near, full capacity, so some expansion of these U.S.-based laboratories is desirable. In the eastern Arctic, U.S. biological research has traditionally been supported through an international agreement with Denmark due to the lack of U.S.-based facilities in Greenland. Establishment of a U.S. Arctic laboratory at Thule or negotiation of an agreement to allow U.S. polar biologists access to Svalbard would provide new opportunities to study northern polar ecosystems.

Due to the difficulties in conducting work during the dark, extremely harsh, polar winter, a substantial fraction of research activity in polar regions is restricted to the warmer parts of the year when the sun is above the horizon. Thus, large gaps in our understanding of organism and ecosystem function exist because processes affecting life in the polar environment occur year-round. Year-round access to terrestrial and marine facilities will not only yield new scientific insights into natural systems but also allow greater flexibility for a broad range of scientists to participate directly in field research. The opportunities provided by winter access could also encourage new participants to enter polar research.

Given the scientific impetus for year-round sample collection and analysis, a base-funded and staffed repository for frozen samples of polar organisms is needed. A sample repository would ensure the proper archiving and curation of samples, ensure the provenance of samples submitted for deposition, and provide accessibility to samples from polar organisms to the broader community of biologists.

*Integration of research activities.* Integration and synthesis of knowledge on the genomes, physiologies, and biochemistries of polar organisms, and the biogeochemical and physical characteristics of polar ecosystems, is an important challenge that must be addressed if polar biology is to realize its full potential. Integration requires creating possibilities for teams of scientists that work within a particular habitat to gather and share information and techniques. Likewise, researchers doing similar research in Arctic and Antarctic ecosystems must be encouraged to share information and insights. Programs such as the Arctic System Science Program serve to help unite the Arctic scientific community both within the United States and internationally. Conferences and workshops can be used to facilitate communication among scientists willing to cross biological disciplines and scales and to develop systemic understanding of the organism, habitat, or region under study. This kind of integration of knowledge will accelerate the infusion of genomics and other techniques into polar biology. Field courses and postdoctoral fellowships should be designed to encourage nonpolar scientists with relevant expertise to pursue studies in the Arctic and Antarctic and to collaborate with polar scientists.

*Increasing the flow of information to nonspecialists.* In addition to increasing interactions between polar biologists and the broader community of biological scientists, continued efforts should be made to enhance the flow of information about polar biology to a wider audience because polar ecosystems play an important role in global-scale phenomena. Thus, what happens to organisms in polar ecosystems may have implications for biological processes in other terrestrial and aquatic ecosystems. Some potential strategies and venues for increasing awareness of polar biology and disseminating new discoveries to a wider audience include the following:

- Coverage of polar topics in textbooks and curricula should be expanded.
- Modern educational technology, such as real-time distance learning, should be used to bring students into close contact with polar biology.
- Additional strategies should be developed for bringing teachers and students into the field.
- Web sites should be developed that provide attractive, informative, and up-to-date information to new audiences.
- Polar scientists should be encouraged to be proactive communicators of discoveries to the media.

Educational and outreach activities in the Arctic should also include the indigenous communities that are part of the ecosystem. The effort should be two-way, with scientists respecting and learning from the experiences of local residents. Encouraging local communities to contribute to research activities seems a wise approach for communicating what science is being conducted—and why—and for identifying research questions and facilitating the research itself.

## FINDINGS AND RECOMMENDATIONS

### A New Unifying Approach to Polar Biological Research

**Finding 1:** Genome science is an addition to, not a replacement for, other approaches to the study of polar biology. The application of new genomic technologies has the potential to be a unifying paradigm for polar biological sciences. Key opportunities include the following:

- Polar organisms and communities offer unique opportunities to study evolution using genome sciences.
- The use of genomic methods will give insights into the effects of global change on polar biota and biogeochemistry.
- Genome sciences have vast potential for elucidating function in microbial communities.
- Polar genome sciences could make broad contributions to biomedicine and biotechnology (for example, cryopreservation, cryosurgery, and cold-functioning enzymes).
- A polar genome research initiative will provide important new information on the evolution, physiology, and biochemistry of polar organisms. Such information not only enhances our understanding of how polar ecosystems function but also helps our search for life in icy worlds.

**Recommendation 1-1:** The National Science Foundation (NSF) should develop a major new initiative in polar genome sciences that emphasizes collaborative multidisciplinary research and coordinates research efforts. The Polar Genome Science Initiative could facilitate genome analyses of polar organisms and support the relevant research on their physiology, biochemistry, ecosystem function, and biotechnological applications.

**Recommendation 1-2:** A new polar genome initiative should capitalize on data from existing Long-Term Ecological Research and Microbial Observatory sites to take advantage of the long-term datasets and the geographical distribution of these sites. Additional approaches may be taken so that research can be conducted at sites with comparable conditions at both poles. For example, there is currently no marine site in the Arctic.

## Coordination is Essential

**Finding 2:** To facilitate the advancement of polar genome sciences, coordination of research efforts will be required to ensure efficient transfer of technologies, provide guidance to researchers on choosing organisms for genome analyses, and help in the development of new scientific initiatives. Coordination of research efforts should begin with syntheses of the available information, thereby avoiding duplication of research efforts. It should facilitate increased communication among the polar scientists and also with nonpolar scientists who have expertise in genomics and other technological advances applicable to polar studies.

**Recommendation 2:** NSF should form a scientific standing committee to establish priorities and coordinate large-scale efforts for genome-enabled polar science (for example, genome sequencing, transcriptome analysis, and coordinated bioinformatics databases).

## Virtual Genome Science Centers

**Finding 3:** Genomic technologies, both those currently available and those anticipated in the future, are applicable to some of the key questions in polar biology. However, the technical demands of genome science often transcend the resources of any individual researcher.

**Recommendation 3:** NSF should support some mechanism to facilitate gene sequencing and related genomic activities beyond the budget of any individual principal investigator, such as virtual genome science centers. The purpose of the virtual centers would be to provide infrastructure for individual researchers and to facilitate technology transfer among researchers. New infrastructure is not needed, rather some type of coordinating body (e.g., University National Oceanographic Laboratory System, Ocean Drilling Program).

## Enabling Technologies

**Finding 4:** Enabling technologies are critical to the successful application of genomic technologies to polar studies.

**Recommendation 4:** Ancillary technologies such as observatories, ice drilling, remote sensing, mooring and autonomous sensors, and isotope approaches should be developed to support application of genomic technologies to polar studies.

### Increasing Awareness and Education

**Finding 5:** Polar systems play important roles in global-scale phenomena and there is a need for enhanced flow of information about polar biology to a wide audience of scientists, policymakers, and the general public.

**Recommendation 5:** NSF should continue its efforts to make information about polar regions available to teachers, schools, and the public. Short- and long-term plans should be developed for increasing public awareness of polar biology. In the near future, postdoctoral fellowships in polar biology could be set up to encourage young scientists to enter the field. Long-term plans should include continued efforts to incorporate polar biology in college and K-12 curricula.

### Impediments to Integrated Polar Science

**Finding 6:** Impediments to conducting multidisciplinary integrated polar science exist, including administrative, fiscal, and infrastructure issues:

- Coordination among directorates within NSF and coordination among agencies are both essential for advancing polar biology.
- International collaborations are vital for all polar research. Current procedures make the involvement of international scientists in U.S. polar biological projects difficult.
- Attempts to conduct comparative research at both poles can be difficult. Although NSF's Office of Polar Programs support research at both poles, grant applications for Arctic and Antarctic research have to be made to two separate NSF research programs. Research proposals often undergo two reviews and scientists must prepare separate budgets for each proposal.
- Infrastructure for Arctic and Antarctic biology needs improvement. The conduct of molecular research in the polar regions requires specific infrastructure, and there is no high-technology equipment for such work in the Arctic. Development of ice-drilling and clean-sampling technologies in the Antarctic will facilitate research in deep ice and subglacial lakes.

**Recommendation 6-1:** To reach the goal of getting excellent science done as efficiently as possible, NSF should remove impediments to cross-directorate funding. Because integrated polar science often requires interagency cooperation, NSF should lead by example and form partnerships with the National Aeronautics and Space Administration and others as relevant. Memoranda of understanding among directorates within NSF and among funding agencies are one mechanism to facilitate transfer of information and coordination of research.

**Recommendation 6-2:** Establishment of international research partnerships or memoranda of understanding will facilitate and enhance these collaborative efforts. Issues such as stipends, travel, visas, education, ship time, aircraft use, and other logistical issues should be addressed in these memoranda to ensure successful operation of international collaborative polar research.

**Recommendation 6-3:** More information is needed to develop solutions to problems related to conducting bipolar research. NSF should conduct a brief survey of researchers and research groups who would potentially work in both poles to identify impediments and then take steps to address them.

**Recommendation 6-4:** To facilitate integrated, multidisciplinary biological research at both poles, NSF will have to improve biological laboratories and research vessels, and develop ice-drilling resources in the polar regions. Opportunities to allow year-round access to, and operation of, field sites should be pursued.

# 1

# Introduction

## WHAT IS GENOMICS?

Genomics, or genome science, is "the study of the structure, content, and evolution of genomes," including the "analysis of the expression and function of both gene and proteins" (Gibson and Muse, 2002). In this context, genomics encompasses functional genomics (gene and protein expression and function), structural genomics (analysis of the three-dimensional structures of proteins), metabolomics (analysis of the metabolites produced and consumed by a population of cells), and many other "-omics" (e.g., ecogenomics, metagenomics, pharmacogenomics, toxicogenomics). Genome sciences make use of, and are integrated by, the related disciplines of bioinformatics and computational biology. These genomic approaches offer global or near-global overviews of gene lists, and gene and protein expression. Furthermore, genomic profiles enable the exploration of the genetic content of organisms that cannot be studied by classical genetic methods. The definitions of these and other specialized terms will be introduced at first use and are summarized in Box 1-1. The major goal of this report is to suggest how these new genomic technologies can foster increased understanding of polar biology by allowing novel types of studies that heretofore were not possible to conduct in polar settings.

## BOX 1-1
## GLOSSARY OF TERMS

| | |
|---|---|
| AFLP® | A DNA fingerprinting technique that detects DNA restriction fragment length polymorphisms by means of PCR amplification. AFLP is a registered trademark of Keygene, N.V. |
| BACs | Bacterial artificial chromosomes that contain large DNA fragment inserts (up to 300 kilobases [kb]) |
| cDNA | A DNA molecule that is complementary to a messenger RNA (mRNA). |
| Consortium | A natural assemblage of organisms in which each benefits from the others |
| Ecogenomics | Use of genome science techniques to study ecology |
| EST (expressed sequence tag) | The partial sequence of a cDNA (either 5', 3', or both) that tags the cDNA as part of a transcribed gene from a cell line, tissue, organ, or organism. |
| Extremophiles | Organisms that grow optimally under one or more chemical or physical extremes |
| Functional genomics | Global analysis of the function of the genes present in genomes, including their regulated expression and the functions and interactions of the encoded proteins |
| Genome | One haploid set of chromosomes with the genes they contain. |
| Genomics | The study of the structure, content, and evolution of genomes, including the analysis of the expression and function of both genes and proteins. |
| High-throughput mode | Rapid and simultaneous processing of large sample sets. |
| Mesophilic | Organisms that grow best at moderate temperatures of 25-40°C |
| Metabolomics | Global analysis of metabolites and metabolic networks in cells, tissues, and organ systems |
| Metagenome | The sum of all genomes in an environment |
| Microarray | A microscope slide or other solid support on which many distinct cDNAs or DNA oligonucleotides are patterned at high density in an addressable array. Microarrays are interrogated by hybridization to fluorescently labelled DNAs or RNAs to detect the genes that are actively transcribed |

## GEOLOGIC AND CLIMATIC TRENDS THAT INFLUENCED EVOLUTION IN THE POLAR REGIONS

The distinct geologic and climatic histories of the Arctic and Antarctic have created two unique polar ecosystems that share some attributes while differing greatly in others. The Antarctic is a glaciated continent

| | |
|---|---|
| | in cells, tissues, or organs or that are differentially transcribed in response to some experimental treatment. |
| Morphospecies | Species whose identities are determined primarily by morphological characteristics |
| Orthologous | Term used to describe two genes that diverged from a common ancestor after a speciation event |
| PACs | P1 artificial chromosomes that contain large DNA fragment inserts (up to 300 kb) |
| Paralogous | Term used to describe two genes that diverged after a gene duplication event |
| PCR; RTPCR | Polymerase chain reaction; reverse transcription polymerase chain reaction |
| Polynya | Open water areas within sea ice |
| Proteomics | Global analysis of protein structure, function, and expression |
| Psychrophilic | Organisms that have an optimal temperature for growth of $\leq 15°C$ and a maximum growth temperature $<20°C$ |
| Psychrotolerant | Organisms that are tolerant to cold temperatures, but whose optimal temperature for growth is $>15°C$ |
| SNP (single nucleotide polymorphism) | A site in a genome that may be occupied by different nucleotides among individuals of the same species. |
| STS (sequence-tagged site) | Any DNA sequence that has been placed on the physical map of a genome. |
| Syntrophy | A nutritional situation in which two or more organisms combine their metabolic capabilities to catabolize a substance not capable of being catabolized by either one alone |
| Thermophilic | Organisms that grow best at temperatures of 50°C or higher. |
| Transcriptomics | Global analysis of RNA transcription at the cellular, tissue, or organismal levels, often using microarrays |
| YACs | Yeast artificial chromosomes containing very large DNA fragment inserts (>1,000 kb) |

NOTE: The genomic approaches mentioned in this report are typically implemented in high-throughput mode.

surrounded by a cold, often ice-covered ocean, while the Arctic is a cold, ice-dominated ocean surrounded by large, continental landmasses. The geologic and climatic histories that led to these different environments set the stage for the evolution of their respective biotas and disparate ecosystems (Figure 1-1).

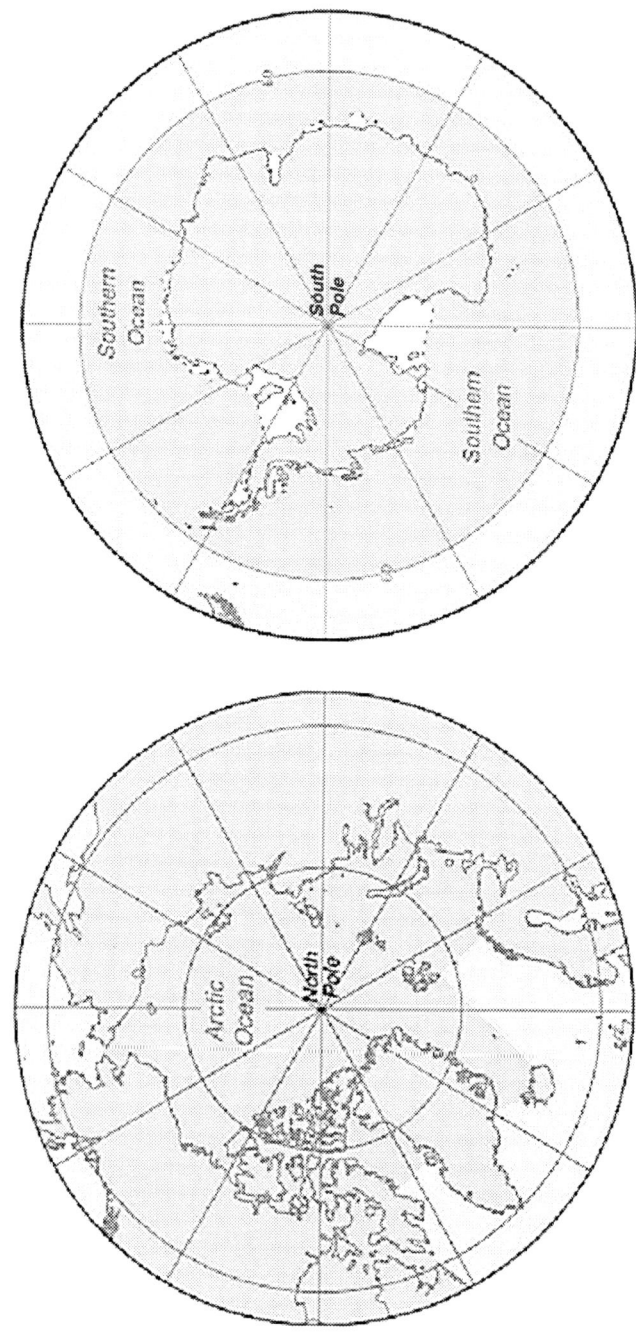

**FIGURE 1-1** Maps of Arctic and Southern Oceans. Available online at <http://www.lib.utexas.edu/maps/polar.html>.

## Antarctica and the Southern Ocean

A globe of the Earth some 250 million years ago would show the continent we know as Antarctica today in the center of the vast supercontinent Pangaea. Over time, major rifting events fragmented the supercontinent until Antarctica developed its present shape. The rifting opened seaways between major oceans and changed the ocean circulation around the Antarctic continent. Throughout this time, Antarctica has remained in the low southern latitudes and has been in a near-polar position for roughly 100 million years (Lawver et al., 1992). Despite this polar position, the climate was initially quite warm. Seas around the continent had bottom-water temperatures ranging from 12 to 16°C (Kennett, 1977, 1982) and supported a complex fish fauna typical of contemporary temperate oceans (Eastman, 1991, 1993), while temperate vegetation flourished on land (Francis, 1999). These temperate climatic conditions ended dramatically when rifting opened crucial oceanic passages, including the Tasmanian Seaway (~35 Ma) and the Drake Passage (~25 Ma), and declining atmospheric carbon dioxide levels combined to trigger profound Antarctic cooling and the onset of rapid glaciation (DeConto and Pollard, 2003).

The East Antarctic continent was likely glaciated for the first time about 34 million years ago (Zachos et al., 2001), but ice extent initially was probably quite variable (Barrett et al., 1987; DeConto and Pollard, 2003). Further cooling shifted East Antarctica into a persistently cold mode (Denton and Hall, 2000) and allowed growth of the more dynamic West Antarctic ice sheet (Alley and Bindschadler, 2001). The general cooling trend over tens of millions of years has been interrupted by important reversals (e.g., Scherer, 2002). Still, overall, the present polar ocean surrounding Antarctica is the most severely and consistently cold marine environment on Earth (DeWitt, 1971; Littlepage, 1965).

Today, the footprint of global change is variable across Antarctica. The Dry Valleys of McMurdo Sound have cooled by 0.7 degree per decade between 1986 and 2000 (Doran et al., 2002). The peninsula also is experiencing significant warming, and several ice shelves on the Antarctic Peninsula have retreated, some reduced to fragments of their original size (Turner et al., 2002). Some of these changes have been dramatic: ice shelves breaking off ice bergs the size of small states or simply disintegrating over the course of weeks, as was the case for the Larsen B ice shelf, where 3,250 km$^2$ of shelf area disintegrated over a 35-day period beginning in January 2002 <http://nsidc.org/iceshelves/larsenb2002/>. The temperature trend for much of the continent remains unresolved due to the paucity of data (Turner et al., 2001).

## The Arctic Ocean and its Surrounding Landmasses

While the geologic framework has had a clear influence on the Antarctic environmental conditions, the tectonic control and the interconnections of major ocean basins in the Arctic are less well defined. When growth of terrestrial ice sheets drew down sea level during the Quaternary, and perhaps earlier, the shallows of the Bering Straits became exposed as a land bridge connecting Asia and North America (Kennett, 1982), blocking direct circulation between the Pacific and Arctic Oceans. Continental motions across the Atlantic Ocean shifted Eurasia and North America apart, contributing to improved communication between the Arctic Ocean and the Atlantic Ocean, while squeezing the Bering Straits and restricting Arctic Ocean communication with the Pacific. Although profound effects on oceanic circulation have resulted, the exact history is still unresolved (Kennett, 1982; NRC, 1991; Aagard et al., 1999).

Extensive glaciation of the Northern Hemisphere post-dated the Antarctic glaciation by over 30 million years. The first major glaciation probably occurred in the late Pliocene (Kennett, 1982) and was certainly occurring by the Pliocene-Pleistocene boundary, about 2.5 million years ago (Shackleton et al., 1984). Since then, multiple cycles of ice sheet accumulation and melting have occurred, on 40,000 year and then 100,000 year cycles, each dramatically altering the biogeography of both terrestrial and marine organisms. These ice-sheet cycles appear to be driven by cycles in Earth's orbit that control the seasonal distribution of the sun's radiation (Clark et al., 1999). The accompanying dramatic shifts included climatic zones displaced as many as 20-30 degrees of latitude and large fluctuations in ocean circulation patterns, sometimes over timescales of just hundreds of years. Dramatic changes in global atmospheric temperature (5-8°C), called Dansgaard-Oeschger cycles, occurred at intervals of 1,000-3,000 years at least within the most recent glacial cycle (Johnsen et al., 1992; GRIP Members; 1993; Grootes et al., 1993; Taylor et al., 1993; Raymo et al., 1998). Moreover, transitions between climatic regimes have been very abrupt; for example, approximately 50 percent of the temperature change associated with the last glacial period occurred in less than a decade (Severinghaus et al., 1998; NRC, 2002).

Today, the sea ice cover of the Arctic Ocean is in transition, losing 3 percent of the area of total ice cover and 7 percent of the area of multiyear ice per decade during recent decades (Johannessen et al., 1999; Kerr, 1999). The decline in sea ice extent is much greater than can be accounted for by natural climate variation (Vinnikov et al., 1999). Recent surface warming and ice thinning in the Arctic may be caused by changes in the state of the Arctic Oscillation mode of atmospheric circulation (Moritz et al., 2002). Global warming due to accumulation of anthropogenic greenhouse gases

# INTRODUCTION

may also be driving the shrinkage, directly or by influencing the Arctic Oscillation. Observations by people indigenous to the Arctic confirm that changes to both the marine and terrestrial ecosystems are occurring more rapidly now than in the past (Krupnik and Jolly, 2002).

## PHYSICAL PARAMETERS THAT SHAPE BIOLOGICAL PROCESSES

The Antarctic and the Arctic present physical parameters that shape their biotic communities and processes, but as a result of their distinct histories some of these parameters are similar and some are quite different.

### Similarities

- Both regions are cold, isolated, and subject to pronounced seasonal cycles of temperature and daylength.
- Glaciers, icebergs, and sea ice profoundly influence the biogeographic distribution of organisms in both regions and provide novel ecological niches for colonization.
- Thermal conditions in both polar regions have served as an effective barrier to colonization by temperate species, although global warming is reducing this barrier.
- Both regions are highly sensitive to anthropogenic impacts, such as chlorofluorocarbon (CFC)-induced ozone holes.

### Differences

- The Southern Ocean has been remarkably cold and stable for at least 8 million years, whereas the Arctic Ocean cooled much more recently (~2.5 Ma). This difference in thermal history may have led to differences in breadth of thermal tolerance by Arctic and Antarctic organisms.
- Arctic surface air temperatures are more variable than those of the Antarctic in annual, seasonal and daily timeframes and often may change by 40-50°C over a few days or on the same date between years. Tolerance of such rapid temperature variability may have driven the adaptive evolution of Arctic organisms in ways that are not experienced by Antarctic species.
- Riverine freshwater and sediment discharge to the Arctic Ocean are substantial, whereas they are virtually nonexistent in the Antarctic.
- Delivery of glacial icebergs, melt water, and till is of greater import to the Southern Ocean.
- The continental shelves of the Arctic Ocean are broad and relatively shallow, whereas those of the Southern Ocean are narrow and deep.

Thus different types of benthic habitats are available in the two polar oceans.

• Surface lakes and permafrost are prominent features of Arctic landmasses, whereas the Antarctic continent is covered by a massive ice sheet that has isolated subterranean lakes.

• Given the predominance of ice in the Antarctic environment, the terrestrial flora and fauna of the Arctic are more diverse.

• The Arctic is home to indigenous human populations. These people have a long and close relationship to the environment and associated biological resources and are affected when these resources change, whether because of natural variability or anthropogenic influences.

## EVOLUTION IN POLAR REGIONS

The genetic structures of northern populations, communities, and species—whether terrestrial or marine—are the "genetic legacy" of rapid Quaternary climate changes (Hewitt, 2000), and the genomes of boreal species are expected to bear the signature of these changes. In the Antarctic, by contrast, the evolution of marine species has been driven by a long period of stable, low temperatures; and the relatively limited terrestrial ecosystems have been shaped by temperatures considerably more severe, but less variable, than those of the north. Indeed, the McMurdo Dry Valleys have been studied as an analogue for potential life on Mars, and subglacial Lake Vostok has been studied as model for possible life on Europa. The key to understanding the mechanisms of biotic evolution in the distinct polar regimes of the north and south lies in analysis of the genomes of organisms from major taxa at the individual, species, population, and community levels of biological organization.

## EXAMPLES OF RESEARCH AREAS THAT COULD BE ADDRESSED WITH GENOMIC TOOLS

The development of sophisticated technologies for genome analysis, as well as other enabling technologies (such as remote sensing, and nanoscale biosensors), promises to revolutionize our understanding of polar organisms, communities, and ecosystems. Areas of research (explored in depth in Chapter 2) that offer potentially valuable opportunities include:

• *Polar ecosystems and global warming.* Climate modeling and direct experimental measurement indicate that environmental change, including warming, will be most extreme in the polar regions. New genetic and genomic technologies, such as transcriptional profiling using microarrays and protein turnover studies via two-dimensional electrophoresis and mass spectrometry, can be leveraged to understand the impact of such change on individual species and on community structure.

- *Ecological impact of ultraviolet radiation.* Due to the anthropogenic ozone hole that forms over Antarctica each austral spring, terrestrial and marine organisms of the photosphere experience levels of ultraviolet radiation that are much higher than those that were present during the evolution of these species. What are the impacts of this exposure on organismal fitness and community structure? Microarrays, proteomics, and metabolomics can be brought to bear to yield quantitative indicators of ecological impact.

- *Evolutionary mechanisms of adaptation to extreme environmental conditions.* What molecular, biochemical, and physiological mechanisms enable polar organisms to survive, grow, reproduce, and indeed thrive, under extreme cold conditions? Because the Arctic and Antarctic regions underwent glaciation during different geological epochs (Pleistocene and Miocene, respectively), comparison of adaptations in ecotypically equivalent boreal and austral taxa will provide important insights into convergent and divergent evolutionary adaptation. Other major environmental variables that have influenced evolution in polar regions include the extreme variability of annual light cycles and the dry conditions of the Antarctic continent.

- *Systematics of polar organisms.* In many instances, the phylogenetic relationships of polar organisms are poorly understood. Total-evidence phylogenetics, which incorporates molecular, cytological, and morphological character sets, can be applied to resolve evolutionary ambiguities and enhance our understanding of the origin and radiation of key taxonomic groups. Analysis of whole genomes will greatly facilitate systematic studies of polar organisms.

- *Gene flow.* Measurement of gene flow between populations is critical to understanding evolutionary speciation. Allele-specific microarray technology can be employed to determine the effect of gene flow between populations on the rates and patterns of speciation in polar regions.

- *Polar regions as extraterrestrial analogues.* The cryptoendolithic and lake-dwelling organisms of the McMurdo Dry Valleys have long been recognized as potential analogues of life (if any) on Mars, just as permafrost formations in the Arctic provide useful frozen habitat analogues. Similarly, the long-isolated (~20 million years), microbial communities of Lake Vostok in Antarctica and the severely chilled microorganisms in winter Arctic sea ice might serve as models for evaluating the potential for life on Europa. Genome analyses of these organisms will provide us with an understanding of their origins and of genetic traits that might be expected in extraterrestrial life.

- *Polar biotechnology.* The uniquely cold-adapted enzymes of polar organisms provide numerous opportunities for biotechnological development. Proteases that function at temperatures near 0°C are already important for food processing and for cold-water detergent formulations. One can envision that enzymes from polar organisms will have numerous commercial applications where maintenance of low temperature is required. Molecules that protect polar organisms against damage from freezing also have important biotechnological applications.

# 2

# Important Questions in Polar Biology

*Unplanned natural experiments create ecological communities that we would never have dreamed of creating....* (Diamond, 2001)

Polar regions present biological phenomena that strikingly illustrate the truth of Jared Diamond's statement. Who could have predicted that study of polar ecosystems would reveal fishes that, unique among vertebrates, lack red blood cells; hibernating mammals whose body temperatures plummet below 0°C in winter; algae, living within ice- and quartz-containing rocks, that may be metabolically active for only hours each year; fishes whose blood remains in the liquid state at subzero temperatures because of the presence of novel biological antifreeze proteins; and large subglacial lakes, isolated from the rest of the biosphere for many millions of years, that may hold a variety of "ancient" forms of life? The fascination that polar ecosystems hold for scientists thus is not difficult to understand. The "novel" or "exotic" nature of many polar organisms cannot fail to spike the curiosity of any biologist interested in how organisms "work" and how they have evolved in the extremes posed by high latitudes.

Polar researchers have a long heritage of contributing biological knowledge—from the jack-of-all disciplines natural scientists who accompanied the great polar explorers to the cutting-edge researchers supported today by the National Science Foundation (NSF) and others. In addition to studying polar organisms because they are inherently fascinating, there are other compelling reasons to expand our nation's efforts in polar biological research. One is to increase our understanding of fundamental biological principles that are common to most, if not all, organisms. Analysis of life in extreme environments often provides a unique perspective on the fundamental characteristics of living processes present in most species. The mechanisms by which different biological processes adapt to environmental extremes (e.g., low temperatures and dichotomous light/dark cycles) can teach us a great deal about the basic characteristics of these systems, for example, by showing how variation in structure of a macromolecule leads to alteration in its function. Polar organisms thus offer powerful study systems for elucidating the fundamental properties of cellular design and the ways in which evolutionary change in the cell adapts organisms to their environments.

Another compelling reason for intensifying our study of polar ecosystems is that they are likely to be among the ecosystems that are most strongly affected by global change. Therefore, if we are to predict how global change—for example, increases in environmental temperature or ultraviolet (UV) light levels—will affect polar ecosystems, we must characterize more fully the environmental impacts of these changes on polar organisms at all levels of biological organization: ecology and physiology to biochemistry and molecular biology. Furthermore, because the potential effects of global change on polar ecosystems may be severe, the implications for people living at high latitudes also have to be addressed. The more fully we understand the effects of global change on ecosystems, the more prepared we will be to predict—and address effectively—these ecological changes and their societal impacts.

In summary, polar ecosystems offer to biologists of all disciplines advantageous study systems for analyzing a wide range of important questions, many of which can now be addressed with the powerful "tool kit" offered by genome sciences in addition to other enabling technologies. This report presents a range of examples of such questions and offers suggestions about how the new technology might be implemented most effectively to study these increasingly important issues.

## EVOLUTION AND BIODIVERSITY OF POLAR ORGANISMS

### Cold Earth: Hotbed of Evolution?

The rapid onset of extreme conditions in the insular polar marine ecosystems has certainly driven the evolution of their biotas. The best documented example of rapid speciation is found in Antarctic fishes of the perciform suborder Notothenioidei (Eastman, 2000; Eastman and McCune, 2000). It is likely that other major taxa have speciated at comparable rates in these "hot beds" of evolutionary change. The Antarctic fish fauna lack the higher taxonomic diversity typical of other inshore marine habitats. The ancestral notothenioid probably arose as a sluggish, bottom-dwelling perciform species that evolved some 40 million to 60 million years ago in the then-temperate shelf waters of the Antarctic continent (DeWitt, 1971; Eastman, 1991, 1993). The grounding of the ice sheet on the continental shelf and changing trophic conditions eliminated the taxonomically diverse late Eocene fauna and initiated the original diversification of notothenioids. On the high Antarctic shelf, notothenioids today dominate the fauna in terms of species diversity, abundance, and biomass, the latter two at levels of 90-95 percent.

In a habitat with few other fishes, notothenioids underwent a rapid phyletic diversification directed away from the ancestral benthic habitat toward pelagic or partially pelagic zooplanktivory and piscivory (see Plate 1; Eastman, 1993). The diversification of notothenioids centered on the alteration of buoyancy. Although they lack swim bladders, some species lowered density to neutral buoyancy through a combination of reduced skeletal mineralization and increased lipid deposition. In the dominant family Nototheniidae, about 50 percent of the Antarctic species inhabit the water column rather than the ancestral benthic habitat. Referred to as pelagization, this evolutionary tailoring of morphology for life in the water column is the hallmark of the notothenioid radiation and has arisen independently several times in different clades (Eastman, 1999). The notothenioid diversification has produced different life history or ecological types similar in magnitude to those displayed by taxonomically unrelated shelf fishes elsewhere in the world. This is unique, and on the basis of habitat dominance and ecological diversification, notothenioids constitutes one of the few examples of a species flock of marine fishes (Eastman, 2000; Eastman and McCune, 2000).

How rapidly did the notothenioid clades speciate? In short, very rapidly. Diversification within the suborder occurred during the mid-Miocene ~5-14 Ma (Bargelloni et al., 1994; Chen et al., 1997a, 1998). Based on this time span for divergence, Eastman and McCune (2000) have calculated that the average time for speciation for 95 notothenioid species was 0.76 million to 2.1 million years, which is similar to estimates for speciation time in the rapidly evolving Lake Tanganyika cichlid flock (Martens, 1997; McCune, 1997). Though polar oceans are cold, they can be "hot spots" of evolution.

*Key Questions*

Given the distinct glacial histories of the Arctic and the Antarctic, the following questions may be asked:

- Do any of the groups of fishes in the Arctic constitute a species flock?
- Are the adaptations of Arctic fishes to freezing conditions similar to, or different from, those of the Antarctic notothenioids?

### How Has Evolution in the Polar Regions Shaped the Genomes of Organisms?

A question of fundamental importance across all biological disciplines asks what types of genetic information are needed to allow organisms to adapt to the abiotic (physical and chemical) features of their environments. This general question must be considered in the context of two different time frames: (1) *long-term evolutionary processes* in which the genetic repertoire of the organism is modified in ways that better adapt the organism to its environment and (2) *shorter-term events* referred to as acclimations that occur within the lifetime of an individual organism, in which the phenotype is modified through differential expression of the organism's genetic information.

An important issue in the investigation of adaptation to abiotic factors concerns the genetic differences between organisms that tolerate wide ranges of different environmental conditions, eurytolerant species, and those that are only narrowly tolerant of environmental change, stenotolerant species. In light of global climate change, it has become of more than purely academic interest to identify the types of genetic mechanisms that provide organisms with the abilities to adapt to environmental change and, conversely, to understand what types of genetic limitations exist in stenotolerant organisms, notably stenothermal organisms that possess very limited abilities to tolerate and acclimate to changes in temperature.

Polar species, especially aquatic ectotherms ("cold-blooded" species), offer promising study systems for addressing questions about the genetic requirements for coping with environmental change. Because they evolved in highly stable environments, some polar species may be among the most stenotolerant organisms in the biosphere. For instance, Antarctic notothenioid fishes are the most stenothermal animals known; they die of heat death at temperatures above 4°C (Somero and DeVries, 1967), and their tolerance of elevated temperatures cannot be increased through long-term laboratory acclimation (Hofmann et al., 2000). The stenothermal character of these fishes is likely due in part to the loss from their genomes of information that encodes proteins that play crucial roles in the response of more "eury-" species to environmental change. A striking example of the loss of ability to adapt to temperature is the apparent loss of the heat-shock response in Antarctic notothenioid fishes. The heat-shock response is the induction of a family of proteins known as heat-shock proteins that function to protect the cell from heat-induced damage to proteins. As part of a larger family of proteins known as molecular chaperones, the heat-shock proteins prevent aggregation of heat-damaged proteins and assist in the refolding of damaged proteins into their natural, functional states. The heat-shock response is generally regarded as a property of all species, yet this "ubiquitous" response could not be detected in Antarctic notothenioids (Hofmann et al., 2000). The message of this study is that Antarctic notothenioids are genetically compromised in their abilities to acclimate to rising water temperatures. Other recent studies have shown that genes encoding the oxygen transport proteins hemoglobin and myoglobin have become dysfunctional in certain Antarctic notothenioids, the icefishes (see Plate 2; family Channichthyidae [Cocca et al., 1995; Sidell et al., 1997; Zhao et al., 1998]). These are the only vertebrates known to lack oxygen-binding transport proteins (Plate 3, Figure 2-1). Losses of the heat-shock response and oxygen transport proteins during the approximately 15 million years of notothenioid evolution at near-freezing temperatures (Clarke and Johnston, 1996) may reflect the absence of a need for these physiological capacities in cold, thermally stable, and oxygen-

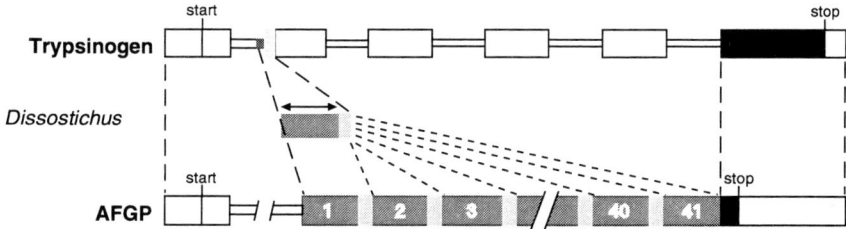

FIGURE 2-1 Structures of the genes that encode trypsinogen and the antifreeze glycoprotein of an Antarctic notothenioid fish (*Dissostichus mawsoni*). Exons are denoted by thick boxes and introns by thin boxes. Gene segments filled with vertical lines are untranslated (regions to the left of the start codon and to the right of the stop codon). Gene segments filled with a checkered pattern indicate signal peptides. The AFGP gene of the Antarctic notothenioid arose from a highly duplicated portion of the trypsinogen gene that comprises parts of the first intron and the second exon. The double-headed arrow shown in the center part of the figure above a single AFGP-encoding segment denotes the expansion of a sequence element present in the trypsinogen gene that has given rise to the canonical AFGP repeat unit. The AFGP-encoding regions of the gene are given numbers to indicate that the gene has 41 AFGP-encoding regions. The lightly shaded regions between the darkly shaded AFGP-coding regions represent proteolytic sites. SOURCE: Figure modified after Logsdon and Doolittle (1997), based on data in Chen et al. (1997a).

rich waters. Mutations in the genes encoding these proteins have led to the loss of physiological capacities normally viewed as "essential to life" and thus carry no evolutionary disadvantage. No doubt other protein-encoding genes and the regulatory networks that govern their expression have also been lost during evolution in highly stable polar environments.

Loss of physiological abilities to cope with increases in temperature characterizes invertebrate species as well as fishes (see Pörtner, 2002), suggesting that stenothermy may be a widespread characteristic of all taxa of polar organisms in both Antarctic and Arctic oceans. However, the loss of abilities to cope with increases in temperature may differ between organisms in the Antarctic and the Arctic Oceans. As discussed in the introductory section of this report, the Antarctic Ocean has had low, stable temperatures for a much longer period than has the Arctic Ocean. Therefore, organisms from Antarctic waters may have lost appreciably more of their abilities to adjust to increased temperatures compared to Arctic species, such that Antarctic marine species may be more susceptible to the effects of global climate change than Arctic species. For terrestrial species, the Arctic environment again has a much wider range of temperatures, so capacities for acclimation would be expected to be greater for Arctic than Antarctic species. However, we know little about the relative abilities of organisms from the two polar regions to acclimate, and the mechanisms of acclimation remain obscure for both groups.

The above discussion of capacities for responding to environmental change prompts a number of lines of inquiry, all of which can reasonably be expected to yield to genomic approaches in the near future.

- *First, how does the repertoire of genetic information change during evolution in highly stable environments, compared to evolution in environments that confront organisms with wide and often rapid changes in key variables such as temperature and oxygen availability?* Does evolution in a stable environment permit loss of genes whose products cease to be needed—as the data on oxygen transport proteins and the heat-shock responses of notothenioid fishes suggest? How widespread among taxa is depletion of the "genetic tool box" and how does the rate at which these genetic tools are lost differ between Antarctic and Arctic species? How do different taxa, including prokaryotes and eukaryotes, compare in terms of loss of genetic information? Does loss of genetic information hamper organisms' abilities to respond to environmental change, for instance, to increases in temperature? Knowing what has been lost in "steno-" species may enable us to predict how difficult it will be for them to cope with climate change. Will species that have lost abilities to acclimate to higher temperatures and to extract oxygen from warmer waters face extinction as the oceans increase in temperature? Which species are most vulnerable?

- *Second, what* new *types of genetic information are needed to permit organisms to cope with polar conditions—and how is this new information generated from preexisting "raw material" in the genome?* Here, the paradigm is the set of genes that encode antifreeze proteins and antifreeze glycoproteins (AFGPs) (see Plate 3; Figure 2-1). These genes have originated multiple times, from several preexisting genes in Antarctic and Arctic fishes (Chen et al., 1997a,b; Fletcher et al., 2001). In the case of Antarctic notothenioid fishes, a gene encoding the proteolytic enzyme trypsinogen has served as the raw material for generation of the gene encoding AFGPs (Figure 2-1; [Chen et al., 1997a]). The notothenioid antifreeze gene is seen to originate from part of a noncoding intron and part of a coding exon of the trypsinogen gene. There had been a massive repetition of the ninenucleotide sequence encoding the canonical antifreeze tripeptide, alanine-alanine-threonine. In the mature glycoprotein antifreezes, galactosyl-N-acetylgalactosamine residues are attached to the threonine. Interestingly, in the Arctic fishes that possess glycoprotein antifreezes with this same primary sequence and carbohydrate composition, a different—yet unidentified—gene, has been recruited as raw material for the antifreeze (Chen et al., 1997b). Several other genes have been recruited for protein antifreezes in fishes. The phenomenon of parallel convergent evolution has been discovered in the study of protein antifreeze-encoding genes in some Arctic fishes (Fletcher et al., 2001). Two species have independently developed antifreeze-encoding genes by modifying genes that code for C-type lectins, proteins that bind carbohydrates.

Antifreeze genes offer potentials for the study of temperature-regulated gene expression. Although antifreeze genes are constitutively expressed in Antarctic notothenioids, antifreeze production in Arctic fish is seasonal and is regulated through a complex hierarchy of regulatory steps involving hormonal signals (Fletcher et al., 2001). Understanding these regulatory cascades will add to our understanding of how gene expression in vertebrates is regulated in response to environmental change.

Macromolecular antifreezes and ice-nucleating agents in terrestrial invertebrates, including insects, spiders, mites, nematodes, and many other organisms, especially sea-ice algae and other psychrophilic microorganisms, also merit additional study (Wharton and Worland, 1998). Intracellular freezing and survival have been demonstrated only in the intact nematode roundworm *Panagrolaimus davidi*, which resides in mosses of glacial meltstreams. The mechanism that nematodes use to resist freezing is similar to the mechanism they use to resist desiccation, where the organisms enter a state of suspended animation known as anhydrobiosis (Browne et al., 2002). The structural properties of antifreezes, the potentiation of their action by ice-nucleating agents, the regulation of production of antifreezes and ice-nucleators, and the process of anhydrobiosis will all benefit from study using tools of genome sciences. Little is known about the genes that have been recruited to fabricate invertebrate antifreezes, so this topic is a frontier for future study.

Are there other instances that genes have been recruited for a new function, in order to allow organisms to adapt to polar conditions? What lessons about molecular evolution can be learned from studying the generation of new genetic capacities in polar organisms? As shown by the seasonal production of antifreezes in Arctic fish, gene regulatory systems have evolved to maintain efficient gene expression in the cold (Fletcher et al., 2001). Thus, studies of the genomes of polar species may provide new insights into the ways in which shifts in environmental conditions such as ambient temperature are transduced into alterations in patterns of gene transcription.

• *Third, what are the genetic mechanisms that cause the genomic changes that lead to rapid evolution in polar environments?* Traditionally, genomes have been regarded as relatively stable entities, undergoing mutations at a rate of $10^{-9}$ nucleotides per year (Kazazian and Goodier, 2002). There are, however, several mechanisms that promote genomic instability. Expansion of short repeats (e.g., trinucleotide repeats) and large-scale deletions, duplications, and inversions of several megabases are recognized as the causes of ~40 human diseases (Cummings and Zoghbi, 2000; Emanuel and Shaikh, 2001). Furthermore, transposable elements (DNA transposons and retrotransposons) are major components of the human genome. For example, the human L1 LINE (long interspersed nuclear element) retrotransposon, which comprises 17 percent of human DNA (Lander et al., 2001), moves about the genome by making RNA copies of itself, reverse-transcribing the L1 RNA into DNA, and then integrating the new copies at other genomic sites. Although most L1 sequences have lost their ability to transpose, some 60-100 copies remain active in the human genome and are able to produce deletions, duplications, and inversions by multiple mechanisms (Kazazian and Goodier, 2002). Recently, Gilbert et al. (2002) and Symer et al. (2002) have shown that L1 transposition in human cell lines produce genomic changes, principally large deletions, in ~10 percent of new insertions. If L1 transposition events occur frequently (estimated at 1 in 10-250 humans born [Ostertag and Kazazian, 2001]) and substantial deletions occur in 10 percent of L1 insertions, then retrotransposition may be a major factor underlying genome, and hence organismal, evolution. The importance of insertions and deletions (collectively termed *indels*) to species divergence is supported by Britten (2002), who recently revised his estimate of the sequence identity of human and chimpanzee genomes from 98.5 percent to 95 percent. Of the 5 percent divergence, 3.4 percent was attributable to indels and only 1.4 percent to base substitutions.

Recent evidence suggests that genomic evolution in the notothenioid fishes of the Antarctic may be based in part on repetitive genetic elements. Parker and Detrich (1998) discovered a notothenioid-specific repetitive element (Noto1, ~285 base pairs [bp]) that is present in an α-tubulin gene cluster of *Notothenia coriiceps* and in the trypsinogen gene of *Dissostichus mawsoni*. Furthermore, preliminary results (H.W. Detrich, III, unpublished observations) suggest that notothenioid genomes contain both LINE retrotransposons and the related, but smaller, SINEs (short interspersed nuclear elements), which can be mobilized to move in genomes by the enzymatic activities encoded by LINEs (Kazazian and Goodier, 2002).

As is clear from the information given above, recent exploitation of the tools of genome sciences in the study of polar organisms has led to many appealing examples of novel mechanisms of adaptation and has opened the door to exciting new lines of study. Thus, it is clear that the information obtained to date through application of genome sciences represents but the very "tip of the iceberg" in terms of what knowledge can be obtained through expansive use of genomic methods, including those of genomics, proteomics, and metabolomics, in the arena of polar biology.

High-throughput sequencing strategies make it realistic to begin sequencing the entire genomes of polar species (see Chapter 3). Sequencing the genome of a notothenioid fish—for instance a "white-blooded" icefish (family Channichthyidae)—might reveal the types of losses that occur during evolution in cold, thermally stable, and oxygen-rich waters. The fact that loss of a trait such as myoglobin expression can result from different types of lesions (Sidell et al., 1997)—in some instances the reading frame is disrupted, whereas in other the mRNA for myoglobin is present but is not translated—suggests that genomic analysis of notothenioids could help reveal the regulatory cascades that govern expression of oxygen transport proteins. Likewise, detecting the lesions that account for the inabilities of icefishes to produce red blood cells (erythrocytes) may help elucidate the complete set of events that underlies the production of red blood cells in vertebrates, including humans (H.W. Detrich, III, unpublished observations). A similar logic applies in the case of the heat-shock response, where the absence of heat-shock proteins is observed even though certain of the gene regulatory proteins governing the heat-shock response are still present (G. Hofmann, personal communication). Finding the lesions that account for the absence of a heat-shock response could elucidate universal components of this "ubiquitous" trait. Analysis of the sequences of genomes of polar organisms, coupled with comparison to the genomes of ecotypically similar temperate species, will advance our understanding of the fundamental processes of molecular evolution and the complex, interconnected pathways involved in the ontogeny and function of cells. Much of what we know about cellular function has come from the exploitation of laboratory-generated mutant organisms in which the disruption of a particular trait allows discovery of the underlying genetic basis of the lesion. Polar organisms can be viewed as "naturally occurring mutant forms" that offer outstanding potential for advancing the biological sciences. The unique properties of the genomes of polar organisms can be discovered only if information is available on the genomes of relevant nonpolar species. The latter information is a necessary comparative backdrop for detecting the key differences that distinguish the genomes of polar and nonpolar organisms.

*Key Questions*

- What new types of information are needed in the genomes of polar organisms for their adaptation to polar conditions?
- How rapidly does molecular evolution occur in polar organisms?
- What types of genetic information have been lost during evolution in cold, stable polar waters, and how might this loss of information prevent polar species from adapting to global climate change?

### How Do the Transcriptomes, Proteomes, and Metabolomes of Polar Organisms Compare to Those of Other Species?

Genome sequencing projects with polar species should be complemented by studies of changes that occur in the transcriptome and the proteome in response to environmental change. Gene expression profiling, using DNA microarrays to monitor shifts in transcription, and proteomic methods to analyze changes in protein patterns, could further enhance our knowledge of differences between "steno-" and "eury-" species. DNA microarray procedures are evolving rapidly, thanks to more genomic information per se and to the successful development of microarrays for nonmodel species (Gracey et al., 2001; Pennisi, 2002). The extension of DNA microarray studies to nonmodel species illustrates the potential advances that these new genomically enabled approaches can make to the study of polar organisms, for which gene sequence data remain relatively rare. Genetically well-characterized species provide a sound frame of reference for the design of studies with polar species. Studies of model species with fully sequenced genomes, such as yeast, have shown that a variety of physical and chemical stressors trigger the production of a common set of stress-associated RNAs, as well as RNAs specific to different stressors (Causton et al., 2001; Gasch et al., 2000). DNA microarray analysis of polar organisms could reveal whether transcriptional responses have been reduced in highly "steno-" species. Microarray studies thus could point to the lesions that limit polar organisms' abilities to acclimate to environmental change, thereby further clarifying the threats that environmental change pose to these species.

As the study of myoglobin production in icefishes has demonstrated, lesions in protein production may lie in events downstream from transcription. Thus, analysis of the transcriptome should be paired with analysis of the proteome. Proteomic analysis in polar species will reveal whether the protein phenotype responds to environmental change in parallel with changes in the transcriptome and whether "steno-" and "eury-" species possess different abilities to alter their protein pools during acclimation. Comparative analysis of transcriptomes and proteomes may reveal differences between Arctic and Antarctic species, differences that reflect the distinct time courses of evolution in thermally stable environments in the two polar regions. The abilities of Arctic species to acclimate could be substantially greater than the abilities of taxonomically similar Antarctic species. Broad taxonomic analyses would reveal whether taxa differ in their abilities to acclimate, for example, to adjust to rising temperatures. If some species possess greater acclimation potentials than other species in an ecosystem, environmental change is likely to have sharply differential effects on different organisms.

Although genome sequencing and analyses of transcriptomes and proteomes is providing vast increases in our understanding of biology, an additional level of analysis is essential if the physiological consequences of environmental change are to be understood. This level of analysis is termed "metabolomics" and comprises characterization of the types and concentrations of low-molecular-mass organic metabolites found in cells (the metabolome) (Fell, 2001; Fiehn, 2001; Phelps et al., 2002; Weckwerth and Fiehn, 2002). Knowing what genes are expressed and what messages within the transcriptome are translated into proteins still gives an incomplete picture of an organism's physiological responses to the environment. There is not a one-to-one relationship between the concentration of messenger RNA (mRNA) and the activity of the protein encoded in the message. Nor is there a one-to-one relationship between the concentration of a protein and its rate of catalytic function, because proteins generally have their activities under tight regulation through post-translational modifications and kinetic control by regulatory metabolites. Thus, characterization of the metabolome can be viewed as the definitive way of gauging what is going on metabolically in the cell.

*Key Questions*

- How do the processes of gene regulation compare in polar and nonpolar species? Do polar species manifest reduced capacities to alter gene expression in the face of environmental change?
- How do the sets of proteins found in cells of polar organisms compare with those found in temperate and tropical species?
- Has any simplification of metabolic processes occurred during evolution at low, stable temperatures?

## "-omic" Approaches Portend Progress in Basic and Applied Research

One of the recurring themes of this report is that a fully integrated understanding of polar organisms will be possible by applying appropriate combinations of genomic, proteomic, and metabolomic approaches. This new appreciation of the biology of polar species will contribute critical new information about the "basic" biology of these organisms (for example, their evolution, their biogeography, and their capacities for responding to environmental change) and may also yield "practical" (biotechnological or biomedical) benefits. Microarray, proteomic, and metabolomic approaches could facilitate the detection of pathological responses to environmental change. The changes in the transcriptome, proteome, and metabolome that occur in response to an alteration in temperature or other abiotic factor may comprise not only adaptive changes that "right an environmental wrong" but also changes that denote a significant decrease in the health of the organisms. As comparative studies of transcriptomes, proteomes, and metabolomes of different species increase, the development of accurate indices for gauging the physiological status ("health") of species will be possible. This type of biotechnological advance will be useful for studying polar and nonpolar species and may have great practical importance for gauging the status of natural populations.

Another instance where practical benefits may accrue from "-omics" research concerns the biotechnological utility of small organic molecules found in cold-adapted species. Low-molecular-weight cyroprotectants (e.g., glycerol), which function as colligative antifreezes, are found in many Arctic arthropods and even in some Arctic fishes in winter (Raymond, 1993). Investigation of cryoprotectants across the full spectrum of polar organisms could be followed by analyses of the proteins that are responsible for their biosynthesis and of the genes that encode these proteins. Once identified, these genes might provide a valuable tool for genetic engineering of animal or plant cells to create cell lines (or cultures of unicellular organisms) that resist freezing and other types of damage from exposure to low temperature. Cryopreservation of biological materials, ranging from purified proteins to cells, tissues, and whole organisms, is an area of biology that could reap enormous benefits from investigations of polar organisms that integrate genomic, proteomic, and metabolomic methodologies.

*Key Question*

• Do polar organisms produce unique organic molecules that could be exploited in biotechnological and biomedical contexts, for example, in maintaining living systems at low temperatures?

## Ice Museums: Do We Have a Record of Evolution and Reintroduced Genomes?

**Ancient DNA**

Research showing that DNA is preserved in the remains of ancient organisms has provided new insights into the study of genetic variation through time (Hofreiter et al., 2001; Wayne et al., 1999). Areas of research that can benefit from ancient DNA include systematics, paleoecology, the origin of diseases, and population evolution (Wayne et al., 1999). Early studies using cloning technology (Higuchi et al., 1984) required a large supply of DNA and could not target specific, single-copy genes (Pääbo, 1989). Cloning technology was eventually replaced by the polymerase chain reaction (PCR), which permits the amplification of specific sequences from only a few template molecules (Mullis and Faloona, 1987). Although studies using PCR-based technology published in the late 1980s and early 1990s claimed that they were able to recover DNA from remains more than 1 million years old, as the field of ancient DNA matured it became apparent that authentication standards were not met in these early studies. The power of PCR was at the base of many of these problems because a single contaminating sequence can potentially outcompete ancient degraded and damaged DNA during PCR amplification (e.g., Pääbo, 1989; Willerslev et al., in press). Upon death of an organism, its DNA is normally degraded by endogenous nucleases to poly- and mononucleotides. Other processes such as oxidation and background radiation can further modify the nitrogenous bases and the sugar-phosphate backbone of DNA. Hydrolytic processes such as deamination and depurination also cause breaks in DNA molecules. Hydrolytic damage takes about 100,000 years to destroy all retrievable DNA under physiological salt concentrations, neutral pH, and a temperature of 15°C (Lindahl, 1993). Fortunately, conditions such as rapid desiccation, low temperatures, and high salt concentrations can help maintain DNA structure intact for a much longer period (Hofreiter et al., 2001). Hence, remains of organisms immured in such polar environments as permafrost and ice hold perhaps the greatest potential for preservation and thus arguably the best supply of material for the study of ancient DNA, a contention supported by recent discoveries in Antarctic ice (Priscu et al., 1999b; Karl et al., 1999) and permafrost (Lambert et al., 2002; Vorobyova et al., 1997).

*Key Questions*

- How long can organisms and nucleic acids be preserved in ice?
- Based on sequence information from ancient DNAs, how rapidly has evolution occurred in polar organisms?
- Are polar environments truly reservoirs of paleogenes that can accelerate the evolution of present day species through lateral gene transfer?
- Can genomes preserved in icy environments provide a signature of paleoclimate?

**Permafrost**

*Microorganisms.* Permafrost represents a relatively stable subzero-temperature environment that allows prolonged preservation of cellular material. The existence of bacteria in permafrost was first reported at the end of the nineteenth century, along with the discovery of mammoths in Siberia (Isachenko, 1912; Omelyansky, 1911). Viable bacteria have since been isolated and cultured from 2-million-year-old Siberian permafrost (Gilichinsky et al., 1992; Vishnivetskaya et al., 2000; Vorobyova et al., 1997). Given the limited amplifying power of PCR in early research and its problems associated with ancient DNA studies outlined in the previous section, the original claims that these cells remained viable for millions of years was challenged. Aspartic acid racemization studies showed that cells from ancient permafrost had a biological age of only 25,000 years, not millions of years, implying that the cells may have been metabolically active during past melting of permafrost or that amino acid metabolism occurred at low rates under frozen conditions (Brinton et al., 2002). Although such low metabolic activity could repair macromolecular damage, it could not lead to net growth. Such microorganisms are likely to have interesting tales to tell about life at low temperature and strategies for long-term survival, even if they are only tens of thousands of years old.

*Metazoans.* Animals recovered from arctic permafrost have been used to examine evolutionary relationships among elephants, bears, and penguins (Lambert et al., 2002; Leonard et al., 2000; Yang et al., 1996). For example, permafrost DNA show that mammoths are closely related to recent elephants (Yang et al., 1996). Ancient DNA studies on remains of Adelie penguins collected from the Ross Sea Coast, Antarctica, suggest that rates of evolution of this species are approximately two to seven times higher than implied by previous indirect phylogenetic estimates (Lambert et al., 2002). Lambert et al. (2002) compared their results with the high rate of hypervariable region I (HVRI) mutation recently reported for humans and concluded that a high evolutionary rate of mitochondrial HVRI is more realistic than previous estimates, particularly for intraspecific comparisons and for closely related species. Those studies illustrate how ancient DNA from frozen environments can provide data to measure the tempo of evolution.

Studies of ancient DNA from permafrost have also been used to estimate population size and fluctuations during possible warming events and have led to a better understanding of dynamics of large mammal and human interchange between Siberia and Alaska. Leonard et al. (2000) measured mitochondrial DNA sequence variation in seven permafrost-preserved brown bear specimens (14,000 to 42,000 years old) to provide a direct analysis of population genetics in the late Pleistocene. Their results indicate possible routes for southern migration. Results from evolutionary and demographic studies using DNA from organisms frozen in permafrost contribute to our understanding of how climatic and other environmental changes during the last glaciation have affected life on our planet.

*Key Questions*

- What mechanisms of freezing tolerance do these organisms encode?
- Can microorganisms in this environment reproduce? If so, there is the exciting possibility that over millions of years, selection for novel cold tolerance mechanisms may have occurred.
- How have consortia of microorganisms changed during evolution in the polar regions?

### Glaciers and Ice Caps

The recent identification of microorganisms in freshwater ice (for example, lake and glacier; Castello et al., 1999; Christner et al., 2001; Karl et al., 1999; Priscu et al., 1998, 1999b) has further expanded the field of study for ancient DNA. More than 70 percent of Earth's freshwater exists as ice, most of which is in Antarctica (84 percent; 12.5 million km$^2$) and Greenland (12 percent, 1.8 million km$^2$), with the remainder present on Arctic islands or as temperate glaciers. In total, ice covers about 15 percent (15 million km$^2$) of the surface of our planet, providing a huge repository for organisms.

Priscu and Christner (in press) estimated that Antarctic ice contains $8.8 \times 10^{25}$ prokaryotic cells and that Antarctic subglacial lakes contain another $1.2 \times 10^{25}$ prokaryotic cells. These prokaryotic abundances equate to about $2.44 \times 10^{-3}$ and $0.33 \times 10^{-3}$ petagrams (1 Pg = $10^{15}$ g) of organic carbon, respectively, which approximates the total amount of organic carbon in Earth's combined fresh waters (i.e., lakes and rivers). Paleoclimate studies have accurately dated layers within the ice caps of both Antarctica (Petit et al., 1999) and Greenland (Alley et al., 1997), providing an important stratigraphic age record for these ice repositories. The deepest ice in Antarctica is at least 420,000 years old (Petit et al., 1999) and may be up to 1 million years old (Siegert et al., 2001). This reservoir of ancient organisms, which include fungi, bacteria, and viruses, provides an unexplored frontier for the study of microbial variability over at least four glacial periods. Future studies of these icy systems should integrate biological and paleoclimatic studies, thereby allowing genomic information to be related to the global changes recorded within the ice. Biological studies should be based on the latest genomic technology such as real-time PCR and competitive PCR in concert with such methods as amino acid racemization and pyrolysis gas chromatography-mass spectroscopy (GC-MS) to determine the identity, age, and physiological state of the organisms preserved in ice cores.

A preliminary assessment (Priscu and Christner, in press) of prokaryotic diversity in ice based on ssu rDNA identity revealed phylogenetic relatedness between bacteria recovered from Antarctica ice and bacteria from permanently cold, nonpolar locales. Psychrophilic and psychrotolerant isolates originate from locations ranging from aquatic and marine ecosystems to terrestrial soils and glacial ice, with little in common except that all are permanently cold or frozen. The wide geographic distribution of related species from diverse frozen environments implies that clades of these bacterial genera evolved under cold circumstances and likely possess similar strategies to survive freezing and to remain active at low temperature. Although it is not possible through analysis of a single gene to adequately characterize the phylogenetic affiliation of a bacterium, a polyphasic approach could reveal patterns of conserved inheritance and divergence from a common ancestor or identify parallel evolutionary pathways. Similar to studies of permafrost "ancient" DNA, we must first establish that slow rates of metabolic activities are not taking place at the ice matrix itself. Evidence for microbial activity within the grain boundaries of "solid ice" exists (Deming, 2002; Price, 2000; Sowers, 2001) and requires further study. If microorganisms indeed metabolize and grow at subzero temperatures at ice grain boundaries, a completely new set of selection pressures must be considered. Although a more ephemeral environment relative to glacial ice, wintertime sea ice epitomizes the multiple and inextricably linked pressures of low temperature, high salt, and limited habitat space (Deming, 2002).

*Key Questions*

• What is the evolutionary origin of the organisms present in the polar ice caps and glaciers?
• If the microorganisms present in ice caps and glaciers are metabolically active, do they possess novel metabolic and biochemical pathways?
• What are the similarities and differences among microorganisms in different subzero environments (for example, ice, permafrost, and subglacial lakes)?
• Do ice-bound microorganisms provide the biological seed to subglacial environments such as Lake Vostok?
• How do multiple selection pressures influence evolutionary processes across a spectrum of ice types, from ancient glacial ice to modern sea ice?

**Genome Recycling**

Rogers et al. (in press) have utilized the isolation of fungi, bacteria, and viruses in ice to examine a temporal form of gene flow they term "genome recycling." The premise of their idea is that organisms that have been trapped in ice for hundreds of thousands to millions of years are eventually released when glaciers calve and the ice melts and mixes with contemporary populations. The mixing of ancient and modern genotypes may lead to a change of allele proportions in the population, which may in turn affect mutation rates, fitness, survival, pathogenicity, and other characteristics of the organisms. Genome recycling is dependent on the revival and establishment of organisms once they emerge from the ice sheets and glaciers, migrate to a suitable location, transfer genetic information to extant populations, and are again transported to ice sheets and glaciers via aeolian processes. This sequence can occur only if the organisms exist in sufficient numbers and the genes survive and propagate within extant populations (Rogers and Rogers, 1999). Genome recycling depends on environmental conditions, transport mechanisms, population sizes, and the fitness of alleles. Rogers et al. (in press) estimate that at least $10^{17}$ to $10^{21}$ viable microorganisms (including fungi, bacteria, and viruses) are released annually from environmental ice.

*Key Questions*

• Is there a threat to humans, animals, plants, or microorganisms from pathogens long immured in ice?
• Are there any new gene products that can be extracted from ice-bound microorganisms?

**Subglacial Lakes**

Antarctic subglacial lakes remain one of the last unexplored repositories of genetic information on our planet. More than 100 subglacial lakes have been identified with the largest (~14,000 km$^2$; >1,000 m deep) being Lake Vostok. At least one other lake with a surface area >600 km$^2$ has been identified near Dome C. These lakes have been ice covered and isolated from direct contact with the atmosphere for perhaps 20 million years. Due to its tectonic origin, Lake Vostok probably existed as a lake well before Antarctica became ice covered. The subglacial lakes are of great interest for the unique organisms they potentially harbor and for the biogeochemical processes that must exist to sustain life with no light, cold temperatures (< 0°C), and low nutrient input. Although no samples have been recovered from subglacial lakes, our understanding of the potential for life in subglacial lakes has been improved by modeling the physical and chemical environment that may be expected in the lakes, by analyzing accreted lake ice, and by studying analogous settings elsewhere.

Predictions of the possible forms of life in subglacial lakes has relied primarily on analysis of accretion ice recovered from the Vostok ice core (Karl et al., 1999; Priscu et al., 1999a). The study of the geochemistry of accreted lake ice meltwater has been used to infer water chemistry of Lake Vostok, for example, whether the lake is fresh or saline and whether the water contains free dissolved oxygen. Estimates are now being made of the nutrient and energy sources needed to sustain an indigenous biological assemblage (Siegert et al., 2001, in press). Despite the limitations of using accretion ice to infer lake conditions, it is the only avenue open for research at this time. The geochemical data are important in predicting the trophic state of the lake, the possible density of microorganisms, and the range of organism types that may reside in the lake. Samples from Lake Vostok should include a suite for genomic, proteomic, and metabolomic analyses to address questions regarding evolution, physiology, and the biogeochemical processes in this unexplored ecosystem. Unlike cells immured in permafrost or glacial ice, subglacial lake systems offer a repository of apparently actively metabolizing organisms whose evolution has been driven by a novel set of environmental conditions.

*Key Questions*

- What are the organisms present in subglacial lakes and what is the diversity of life forms? Is the biology viable or fossil? Which organisms are metabolically active?
- What are the redox couples that support life? What are the energy sources and how is energy extracted from the environment? What are the carbon sources that support life in the lakes?
- What are the biota present in the sediment record of subglacial lakes and what is their evolutionary history?

**What Factors Control the Biodiversity of Polar Organisms?**

Factors affecting diversity and speciation (such as identity of parent species, habitat characteristics and type of selection pressure, and genetic exchange with populations of sibling species from higher latitudes) are likely to differ between the Arctic and Antarctic because of the differences in origins of their landmasses, geomorphology, and geographic connectedness of terrestrial and marine environments. When examined across taxa, rates of genetic exchange between populations of related polar organisms are likely to differ depending on dispersal mechanisms or routes of exchange. As a result, speciation and endemism are likely to differ greatly between major categories of organisms, directly affecting the biodiversity of polar ecosystems.

For some groups of organisms, speciation and endemism are easy to detect (for example, polar bears and walruses are found only in the Arctic; emperor penguins and the notothenioid fishes are found only in the Antarctic). Speciation becomes less obvious for organisms that lack distinguishing morphological characteristics or that are pleiomorphic. In these cases, phylogenetic and genomic evidence can be brought to bear on such questions. Furthermore, these tools may provide the only metric of the degree and rate of speciation for organisms that do not fossilize or for which no fossil intermediates are known.

**Plant Diversity**

Arctic plant communities have a relatively high diversity of functional groups (for example, shrubs, sedges, mosses) in close proximity but low species diversity within each group (only a few dominant species). The overall functioning of Arctic ecosystems is sensitive to changes in species within functional groups and to shifts among groups. For example, in wet meadow tundra, methane ($CH_4$) emission is highly sensitive to differences in the sedge community and their individual abilities to transport $CH_4$ out of the soil (Schimel, 1995). An example of sensitivity to variation among functional groups comes from Alaskan upland tundra, where deciduous shrubs, primarily dwarf birch, *Betula nana*, are highly productive in warm dry years, while the tussock-forming sedge *Eriophorum vaginatum* survives better than *Betula nana* in colder, wetter years (Chapin and Shaver, 1985). Those species therefore complement each other and stabilize the overall functioning of the ecosystem under historical climates. However, under warm conditions and high nutrient inputs, *Betula* becomes dominant in the ecosystem, completely changing almost every aspect of ecosystem functioning including carbon storage, nutrient cycling, trace gas emissions, snow retention and overwinter soil temperatures, palatability by herbivores, and the complex interactions of all these factors. The shift from tussock to shrub tundra represents a state change of the entire ecosystem. On the other hand, *Betula* in Scandinavia does not respond in the same way as Alaskan *Betula* (Grellmann, 2002). Thus, the Arctic ecosystem is highly sensitive to shifts in the plant community at all levels, including both growth form, individual species, and possibly even genetic variation within species. Predicting how Arctic ecosystems will respond to environmental change requires a better understanding of the functional roles of different plant species and their specific responses to various environmental features. Many of these questions can be addressed by traditional approaches. However, understanding of physiological and genetic characteristics is necessary to address questions such as the differential behavior of *Betula* (perhaps the classic Arctic plant) in Alaska versus Scandanavia. Hence, genomic tools will have an important role.

In contrast to the Arctic, the Antarctic plant community is low in diversity of functional groups. Indeed, there are only two native vascular or flowering plant species, *Deschampsia antarctica* and *Colobanthus quitensis*. They occur in the Andes of South America, on several subantarctic islands in the Southern Ocean, and along the west coast of the Antarctic Peninsula. They appear to have colonized the peninsula relatively recently (Holocene), although little is known of the spatial or temporal patterns of these colonizations or the genetic diversity of their populations. Their distribution along the Antarctic Peninsula is extremely patchy, consisting of >150 localities on islands and points along the west coast (Komarkova et al., 1985). These observations pose several important questions ranging from those pertaining to classical island biogeography to newer issues associated with genetic diversity that could benefit from genetic methods (for example, amplified fragment length polymorphism [AFLP®], Mueller and Wolfenbarger, 1999; gene-targeted markers, van Tienderen et al., in press).

*Key Questions*

- How will the Arctic plant ecosystem respond to environmental changes, such as temperature and precipitation?
- What confers the differential physiological responses to environmental conditions of *Betula* in Scandinavia versus Alaska?
- How genetically diverse are Antarctic plant populations? Does sexual or asexual reproduction appear dominant? What are the general spatial and temporal patterns of colonization along the peninsula?
- The rapid changes in climate that have occurred over the last half century in certain polar regions, such as the west coast of the Antarctic Peninsula (for example, ozone depletion-enhanced UV-B radiation, warming), raise many questions, including several at the population level that could benefit from genomic approaches. Have environmental changes led to genetic changes in the populations, for instance? Has recent climate change favored selection for certain genotypes?

**Polar Soil Communities**

The last major effort to describe polar microbial communities was part of the Tundra Biome studies of the International Biological Program nearly 30 years ago (Brown et al., 1980; Hobbie, 1980). Our understanding of microbial diversity in other soil environments, such as the McMurdo Dry Valleys of Antarctica, has advanced substantially due to development of molecular approaches to community analysis (Ranjard et al., 2000), particularly approaches based on analysis of ssu rRNA sequences. For example, Frati and colleagues (Frati and Dell'Ampio, 2000; Frati et al., 2000, 2001) have used molecular biological techniques to examine the evolutionary relationships and population genetics of a springtail, *Collembola*, from Dry Valley soils. Courtright et al. (2000) used both nuclear and mitochondrial gene sequences that encode RNA to examine the genetic diversity of a nematode across the soils of the Dry Valley region.

Analyses based on ssu rDNA gene analyses have overcome the limitations of traditional culture-based analysis of the phylogeny and biogeography of bacteria, including those found in soil. Culture-based techniques are estimated to detect only a small fraction (<1 percent) of the bacteria found in most microbial communities (Figure 2-2). According to ssu rDNA analyses of temperate samples, soil environments typically have some of the most diverse microbial communities on the globe. Only one study used a molecular approach to investigate microbial prokaryotic diversity in polar soil (Zhou et al., 1997). Projects within the National Science Foundation's (NSF) Life in Extreme Environments (LexEn), Microbial Observatories, and Long-Term Ecological Research (LTER) pro-

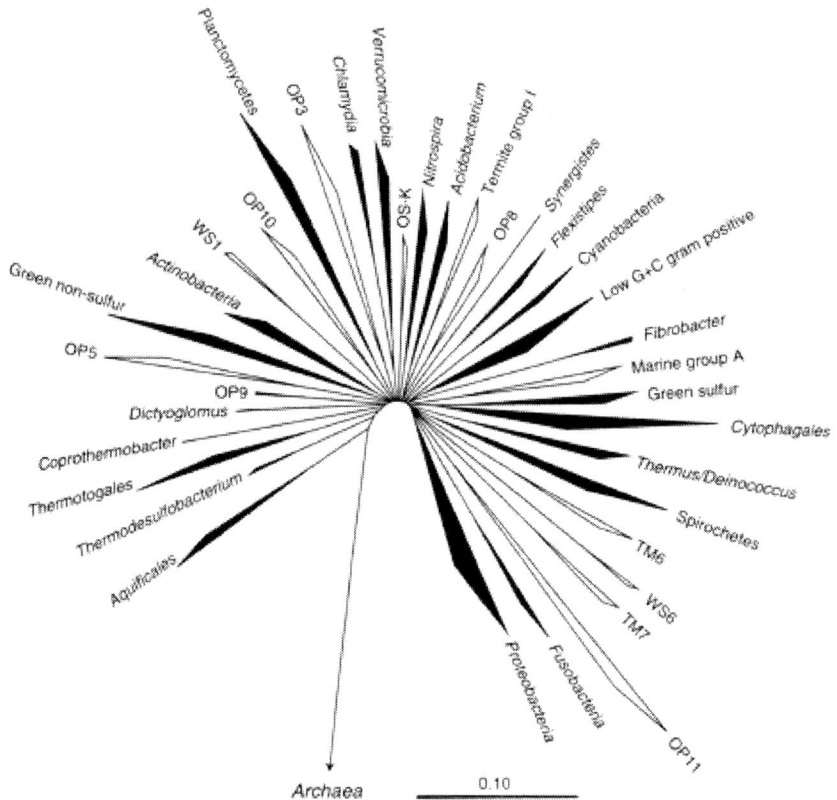

FIGURE 2-2 Evolutionary distance tree of the bacterial domain showing currently recognized divisions and putative (candidate) divisions. The tree was constructed using the ARB software package (with the Lane mask and Olsen rate-corrected neighbor-joining options) and a sequence database modified from Hugenholtz et al. (1998). Division-level groupings of two or more sequences are depicted as wedges. The depth of the wedge reflects the branching depth of the representatives selected for a particular division. Divisions that have cultivated representatives are shown in black; divisions represented only by environmental sequences are shown in outline. The scale bar indicates 0.1 change per nucleotide. The aligned, unmasked datasets used for this figure are available from <http://crab2.berkeley.edu/pacelab/176.htm>.

grams are currently employing molecular techniques ranging from clone libraries of ssu rRNA genes to metagenomic analysis to characterize soil microbial communities in the Arctic tundra (Arctic Tundra LTER); (Broughton, 2002) and boreal forests (Bonanza Creek LTER); (Schloss et al., 2002) and in the Antarctic Dry Valleys. These projects will provide initial information on the composition and diversity of soil microbial communities, including fungi, and will begin relating composition to function. For example, the Bonanza Creek project is focusing on the role of soil microorganisms in phosphorus mobilization.

Polar soil microbial communities contain organisms with cold-adapted metabolic capabilities that may be useful for bioremediation of polluted environments in both polar and temperate regions. Although the potential of microorganisms from polar soils to degrade a range of hydrocarbon pollutants is clear (Aislabie et al., 2000; Braddock et al., 1997; Whyte et al., 1996; Yu et al., 2000), delineation of metabolic activities of polar microbiota has only begun. Hence, characterization of polar soil microbial diversity will help the identification of microorganisms that are effective for various bioremediation activities.

*Key Questions*

- How does soil biodiversity vary over a polar latitudinal gradient?
- What role do polar soil microbiota play in important biogeochemical cycles or in processes such as methane oxidation?
- Do polar microbiota produce cold-adapted enzymes that may have biotechnological or industrial applications?

**Polar Marine Biodiversity**

Some examples of the differences in the composition of polar marine communities have been given above (see discussion of icefish physiology). This section presents additional examples in which insights into the speciation and biodiversity of groups of marine organisms have been gained from the application of genomically enabled techniques to questions of the distribution of organisms and the composition of communities.

One of the most important groups of polar zooplankton is the euphausids, or krill. This group has representatives in aquatic habitats worldwide, but it achieves particular importance in polar environments, particularly the Antarctic, where krill are a mainstay of Antarctic foodwebs. They are also the target of a commercial fishery so that information on stock composition and life history is important for fishery management. Recent work using molecular techniques has improved our understanding of euphausid population biology. For example, Patarnello et al. (1996) have used various genetic markers to show that populations of Antarctic krill from waters south of the Polar Front are genetically distinct from those in the South Atlantic Ocean.

Our understanding of the biodiversity, ecology, biogeography, and physiology of protistan plankton has also advanced as a result of the application of genome-enabled techniques. Gast et al. (2002) are using ssu rDNA libraries to identify small, nondescript, heterotrophic, and phagotrophic protists in Antarctic waters. Recent biogeographical studies of the ubiquitous and important polar phytoplankter *Phaeocystis* have shown that Arctic and Antarctic species are genetically distinct (Medlin et al., 1994; Vaulot et al., 1994). In contrast, Darling et al. (2000) reported that several species (as defined by morphological characteristics, "morphospecies") of foraminifera taken from each polar ocean possessed identical genotypes, indicating rapid genetic exchange between these populations. Intriguingly, the identical genotypes occurred within a morphospecies complex that also contained a range of unrelated genotypes. This finding has significant implications for paleooceanography and climate change research, since distributions of tests from foraminifera morphospecies in sediment cores are widely used to infer environmental conditions at the time the tests were produced. This analysis is predicated on the assumption that a morphospecies is a true species that is uniquely adapting to a specific set of environmental conditions.

Prokaryotes provide a unique challenge to ecologists and biogeographers. Because they are difficult to culture, little is known about speciation, physiological function, or composition of microbial communities found in the two polar oceans. Nevertheless, whether the bacterioplankton species in polar oceans are the same or different is an important question from the standpoint of biogeography, biogeochemistry, and genetic exchange. Recent advances in genome-enabled techniques have advanced our understanding of many aspects of prokaryotic microbiology, ecology, and physiology far more rapidly than they have for protists and other "higher" organisms.

In contrast to metazoa, speciation and polar endemism have been demonstrated for only one group of bacteria, the gas-vacuolate bacteria (Gosink et al., 1997) and one group of cyanobacteria related to *Oscillatoria* (Nadeau et al., 2001). These conclusions were based partially on comparison of ssu ribosomal gene sequences of sibling species, but were subsequently confirmed with taxonomic examinations of cultured organisms. Gas-vacuolate bacteria, which are common in stratified freshwater environments (Walsby, 1994), had not been reported from the marine environment prior to their isolation from Arctic and Antarctic sea ice by Staley and coworkers (Gosink et al., 1993; Irgens et al., 1989; Staley et al., 1989).

Although this phenotype has broad phyletic distribution (Walsby, 1994), Arctic and Antarctic isolates represent distinct endemic species within one genus (Gosink et al., 1997), presumably as a result of the constraints on dispersion imposed by their buoyancy. In contrast, Nadeau et al. (2001) found that psychrotolerant *Oscillatoria* strains isolated from Arctic and Antarctic meltwater ponds were identical, while psychrophilic strains from Antarctica were genetically distinct.

Inferences about speciation and the distribution of uncultured prokaryotes are now being made from ssu gene sequences (Delong et al., 1994; Murray et al., 1998, 1999; Hollibaugh et al., 2002). An example of this kind of analysis is shown in Figure 2-3, where denaturing gradient gel electrophoresis was used to obtain a "fingerprint" of bacterial communities found at different locations in the Arctic Ocean (Bano and Hollibaugh, 2002). The organisms represented by the bands in these fingerprints were identified by cloning and sequencing. However, even though ribosomal gene sequences are useful phylogenetic markers and are the most commonly used "molecular clock," extrapolation of ssu rRNA gene sequence variations to "speciation" may be constrained by lack of information about the rest of an organism's genome.

The shortcomings of the "ssu gene only" approach to understanding speciation have been demonstrated by Béjà et al. (2002b) and Blank et al. (2002). Upon sequencing regions adjacent to ssu genes, Béjà et al. (2002b) found substantial variability among Crenarchaeota that possessed otherwise identical ssu sequences. Furthermore, the DNA fragments analyzed by Béjà et al. (2002b) were sufficiently long to demonstrate differences in gene arrangement among organisms with the same ssu sequence. Although differences in sequence similarity in the spacer region between genes are no surprise (for example, hypervariability of the 16S-23S intergenic spacer region is widely exploited to distinguish between related strains in public health microbiology and microbial ecology [Aakra et al., 1999; Chun et al., 1999; Martinez and Valera, 2000; Tan et al., 2001]), the results of Béjà et al. (2002a) interject a third level of complexity, that of genome order, into this analysis of the distribution of bacteria and the species concept of bacteria.

This example points to the need for polar microbiologists to go beyond the examination of ssu gene sequences to understand the evolution of polar prokaryotes. Understanding the roles that gene order and higher-level variability in microbial genomes play in prokaryote speciation and whether or how this variability is translated into adaptation of prokaryotes to the polar (and other) environments is the current challenge for microbiologists and microbial ecologists. Fortunately, the revolution in genome sequencing is providing a means of tackling these problems. Addressing them will ultimately depend upon generating a significant dataset of

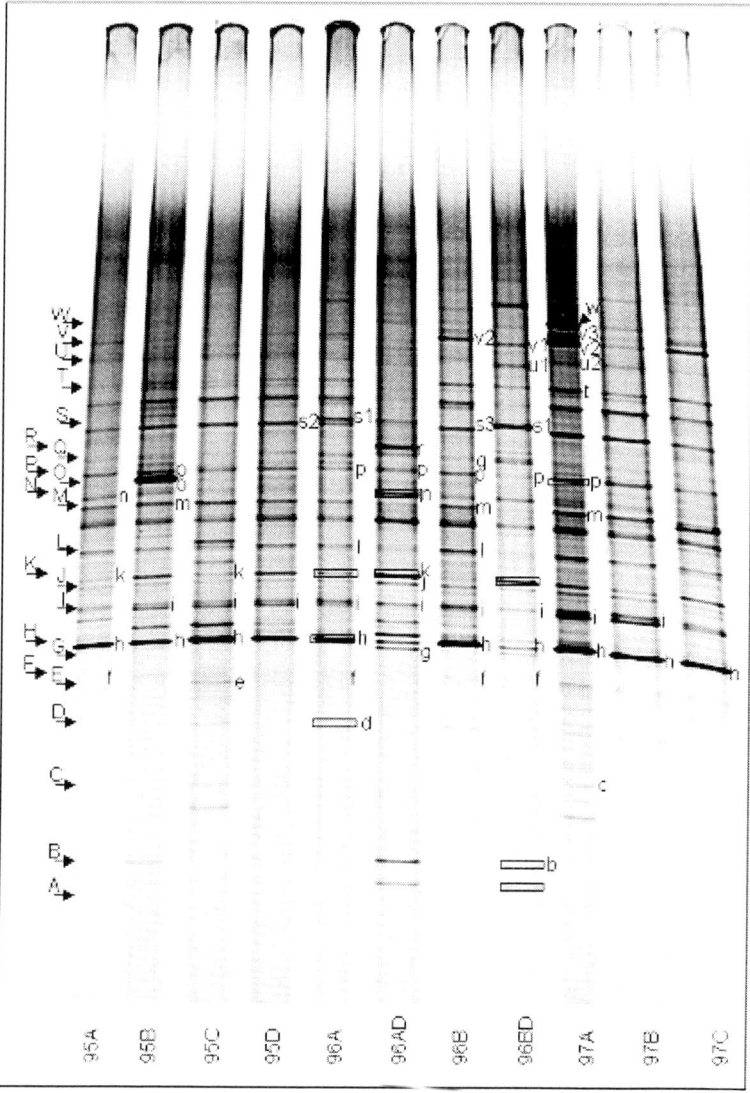

**FIGURE 2-3** Composition of bacterioplankton populations in the Arctic Ocean. PCR was used to amplify portions of ssu rRNA genes from bacteria harvested from water samples. The products of the mixed template amplification were resolved by denaturing gradient gel electrophoresis (DGGE) to show how the composition of the bacteria community varies with depth, time, and location in the Arctic Ocean. Arrows and boxes denote bands that correspond to longer cloned fragments and bands that were excised and sequenced, respectively. SOURCE: Bano and Hollibaugh (2002).

genome sequences from polar prokaryotes for comparative genomic analysis, the application of bioinformatics analysis to the dataset, and the whole sequence of "-omics" techniques to interpret and verify conclusions from these analyses.

One of the most important findings to come from comparative genomics is the fact that a significant number (20-60 percent) of predicted coding sequences in every genome completed to date represent proteins of unknown function. Some of these will likely carry out activities that we are familiar with, others may represent novel cellular functions, and the rest may represent species-specific proteins.

Essentially all of the work to date in microbial genomics has been focused on prokaryotes that can be grown in pure culture or in association with host cells in culture. However, the advancement of genomic technologies makes possible the examination of microbial communities through DNA sequence and microarray analysis without the need for a pure culture as a starting point (Béjà et al., 2000; 2002a; Rondon et al., 2000). In addition, recent progress has been made in developing novel approaches for isolating previously "uncultivable" microorganisms (Kaeberlin et al., 2002; Rappé et al., 2002). Taken together, these continuing advances will likely lead to the discovery of unknown microbial species and unknown functions of microorganisms.

One of the lessons learned from microbial genomics efforts so far is that standard taxonomic methods are not sufficient to understand microbial evolution and relationships, as evidenced by the extraordinary genome plasticity that has been observed, even among closely related species and strains. Although the predominant mode of inheritance appears to be vertical (i.e., from ancestor to progeny), most microbial genomes contain a number of genes that could only have been acquired through horizontal transfer of genes (Koonin et al., 2001). For example, one-quarter of the genome (~450 genes) of the deeply branching bacterial species *Thermotoga maritima* represents genes that were acquired via horizontal gene transfer from a number of Archaeal species (Nelson et al., 1999). Moreover *T. maritima* and many of the potential archaeal donor species were isolated from the same site in Vulcano, Italy. As another example, a comparison between two strains of *Escherichia coli* K12 and the pathogenic O157:H7 revealed that nearly one-third of the genes in these strains are different and that they are scattered throughout the genomes in islands of unique sequence. Approximately 10 percent of the O157 genome appears to have recently been acquired by horizontal gene transfer (Perna et al., 2001).

Several mechanisms for horizontal gene transfer have been identified or postulated, such as transformation (natural competence), conjugation (plasmid transfer), and transduction (phage-mediated gene exchange) (Jain et al., 2002). Insights on horizontal gene transfer to date have been derived from studies of microorganisms that have been grown in pure culture. The relative importance of these processes in nature, the ways these processes influence the efficiency of gene transfer, and the types of genes being transferred are poorly understood. Moreover, it is not clear whether there are any species barriers to horizontal gene transfer or whether all genes can be exchanged between all microorganisms at similar frequencies. Therefore, the refinement of existing approaches to genome analysis is essential for a more detailed understanding of gene transfer.

*Key Questions*

- Do the genomes of polar microorganisms contain uniquely polar features (for example, unique base pair composition)?
- Is there significant genetic exchange between populations of specific microorganisms in polar environments?
- What unique adaptations to their environment do polar organisms display?
- What is the relationship between sea-ice microbial communities and those in the underlying seawater and sediments?
- What controls microbial species succession in polar waters and sea-ice communities?
- Are there unique microorganisms with unique physiological properties in microbial communities?
- Is the frequency of horizontal gene transfer affected by polar conditions (e.g., temperature)?

**Latitudinal Compression and Biodiversity**

Ecosystem gradients provide an intersection of environmental factors leading to novel assemblages of organisms with high genetic diversity and productivity. Polar systems offer a spatial compression of environmental gradients over relatively short latitudinal distances and show greater sensitivity to climate change than middle or low latitudes. This sensitivity is particularly evident during the summer when small temperature changes influence the phase transition of water between liquid and solid. This phase transition has importance for both terrestrial and marine systems (Doran et al., 2002). Polar latitudinal gradients can therefore be used to study the effects of potential changes in regional climate that may or may not be associated with global change and to provide a range of environmental conditions for more fundamental studies.

The importance of the compressed latitudinal gradient across terrestrial and marine ecosystems in Victoria Land has recently become a focus of study by several national programs (see Plate 5); (Berkman and Tipton-

**PLATE 1** Photo of *Dissostichus mawsoni*, commonly known as the Antarctic toothfish. This species is an example of Notothenioid that has evolved neutral buoyancy through reduction of skeletal mineralization and increased lipid deposition. SOURCE: Photo taken by Kevin Hoefling and provided courtesy of Dr. Chris Cheng, University of Illinois at Urbana-Champaign.

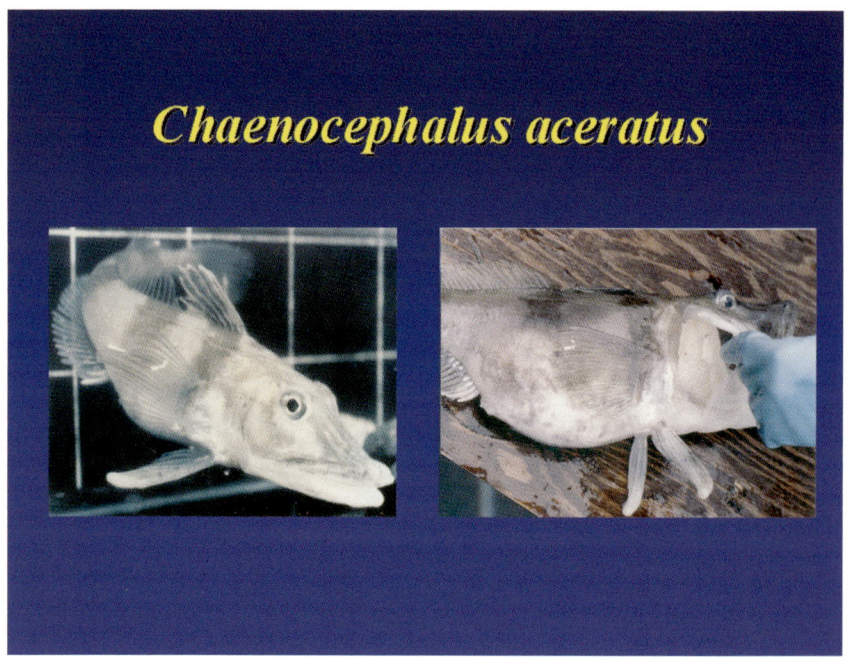

PLATE 2 The icefish *Chaenocephalus aceratus*. (Left) An adult male of ~45-cm total length. The background grid measures 10 x 10 cm. (Right) A living, gravid female. Lifting the operculum reveals the white complexion of the gills due to the absence of red blood cells. In red-blooded relatives, such as *Notothenia coriiceps* (not shown), the gills are a brilliant crimson due to oxygenation of hemoglobin in red cells. SOURCE: Photos by H.W. Detrich, III.


PLATE 3 Globin-related sequences in the genomes of red- and white-blooded Antarctic fishes. (A and B) Southern blots of genomic DNAs from four red-blooded fishes: Gg, the Antarctic humped rocked (*Gobionotothen gibberifrons*); Na, the New Zealand black cod (*Notothenia angustata*); Nc, the Antarctic yellowbelly rockcod (*Notothenia coriiceps*); and Pc, the Antarctic dragonfish (*Parachaenichthys charcoti*); and three white-blooded Antarctic icefishes: Ca, the blackfin icefish (*Chaenocephalus aceratus*); Cg, the mackerel icefish (*Champsocephalus gunnari*); and Cr, the ocellated icefish (*Chionodraco rastrospinosus*) were hybridized to *N. coriiceps* cDNAs for alpha-globin (A) or for beta-globin (B). Note that all fish species have DNA fragments that were recognized by the alpha-globin probes, whereas only the four red-blooded species were positive for beta-globin. These results indicate that the three icefishes have lost the gene for beta-globin. (C) Cartoon depicting the loss of globin genes by the 16 species of the icefish family. Note that *Neopagetopsis ionah* retains a complete, but defective, copy of the alpha-beta globin gene complex, 14 of the 16 icefish species possess a fragment of the alpha-globin

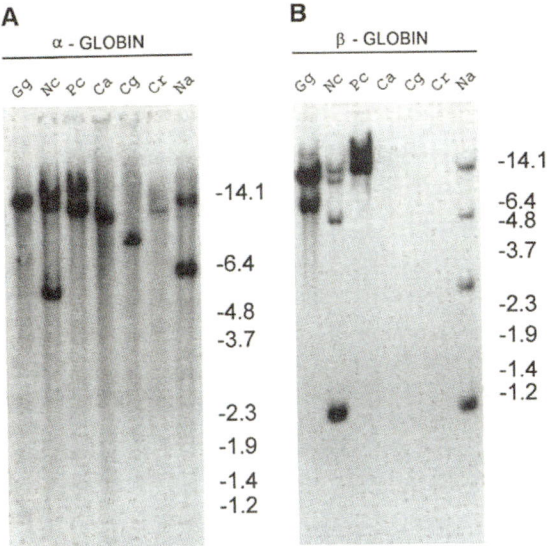

## The Adult Globin Gene Complex of Antarctic Fishes

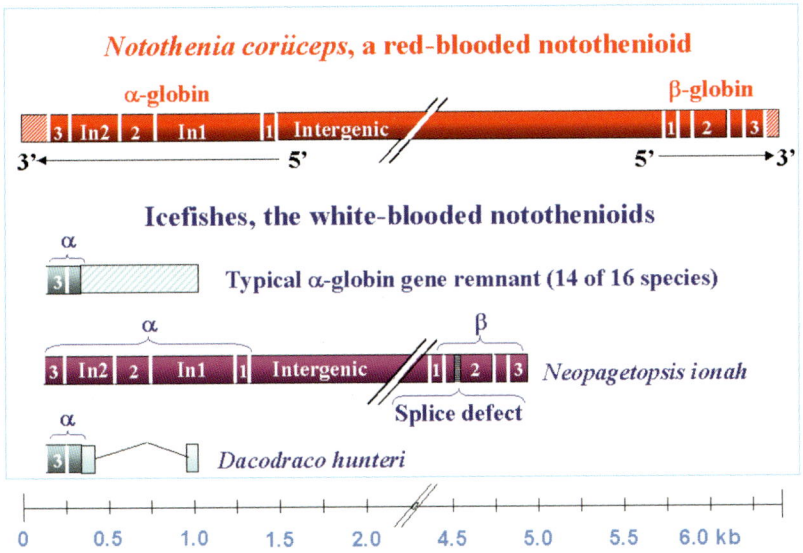

gene only, and the 16th species, *Dacodraco hunteri*, has a further deletion in the alpha-globin gene remnant. These data imply that globin gene loss occurred a minimum of three times during diversification of the hemoglobinless icefishes. SOURCE: A and B, Cocca et al., 1995; C, H.W. Detrich, III unpublished results.

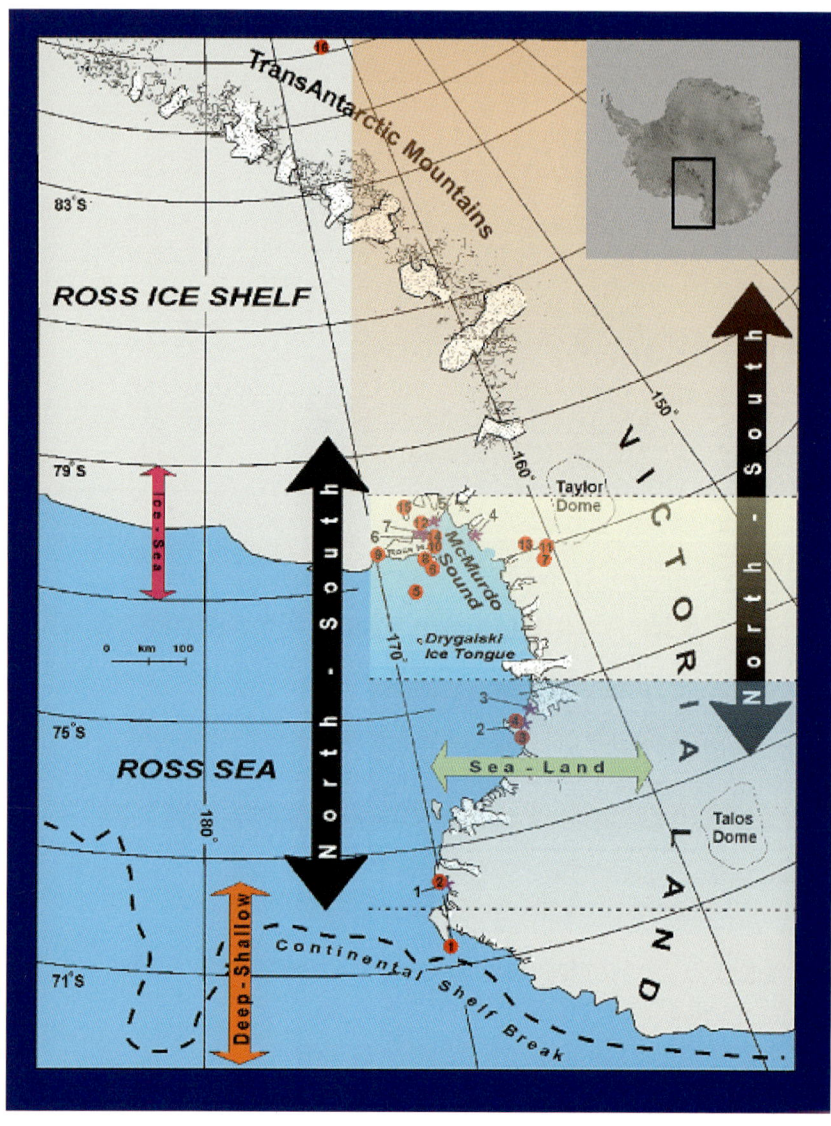

PLATE 4 Map showing the latitudinal and sea-land gradients along the Victoria Land coast being studied by the Italian and New Zealand research programs. Red spots indicate sites of possible sample collection. SOURCE: Berkman and Tipton-Everett, 2001.

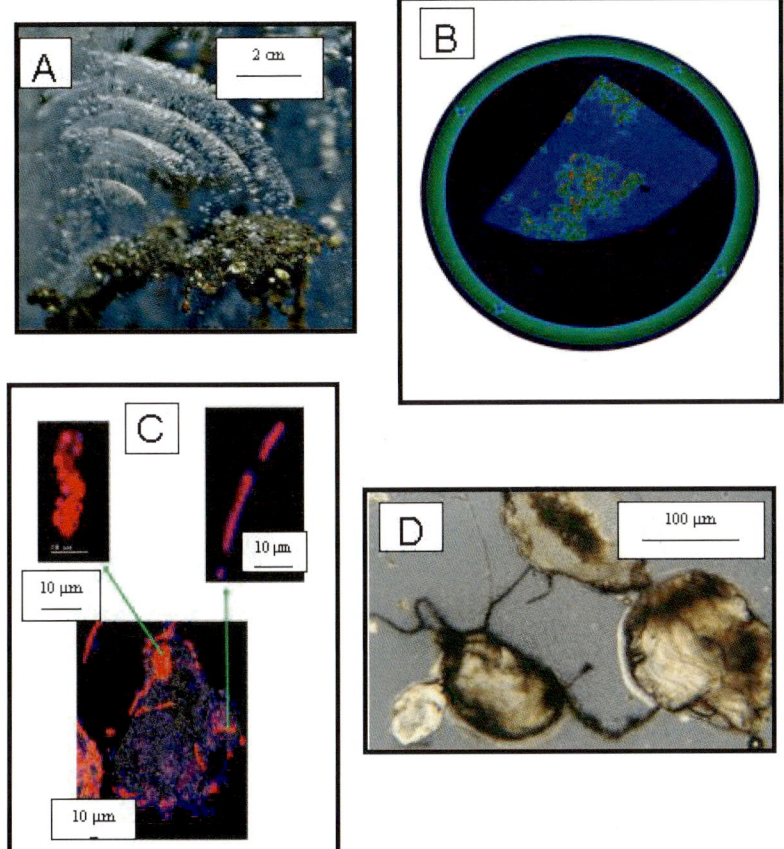

**PLATE 5** (A) Photographic presentation of an aggregate from 2 m beneath the surface of the permanent ice cover of Lake Bonney located in the upper Taylor Valley of the McMurdo Dry Valley system. The photograph was taken from within a 3.5-m deep trench cut into the ice in early September before the formation of liquid water. (B) Computed tomography (CT) scan of a section of ice core from Lake Vida, Victoria Valley (blue = ice, black spheres within the core section = gas bubbles in the ice, red-orange = sediment particles, green = particulate organic matter). The inner diameter of the circular sample chamber (blue-green ring) surrounding the ice core section is 76 mm. (C) Confocal laser photomicrographs showing microorganisms associated with a sediment particle with enlarged views of two species of cyanobacteria (blue = DAPI stained bacteria, red = chlorophyll autofluorescence, gray = sediment particle). (D) $^{14}CO_2$ microautoradiograph of sediment particles bound together by cyanobacterial filaments (dark regions denote sites of active $^{14}C$ accumulation indicative of photosynthetic activity). NOTE: DAPI = 4',6-diamidino-2-phenylindole, dihydrochloride. SOURCE: Modified from Priscu et al., 1998.

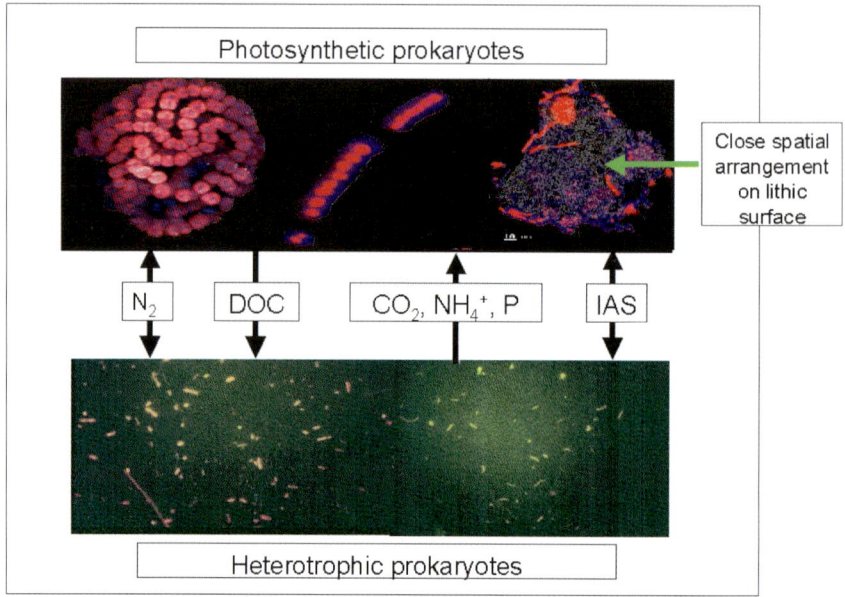

**PLATE 6** Consortial relationships between photosynthetic and heterotrophic prokaryotes found within the ice covers of the McMurdo Dry Valley lakes. NOTE: IAS = ice active substances. $N_2$ indicates that both groups of prokaryotes fix atmospheric nitrogen that is exchanged between them. All of these exchanges take place on micron or smaller scales. Such a consortium is necessary for the survival and proliferation of life in the extreme environment posed by permanent Antarctic lake ice. SOURCE: Priscu et al., in press.

**PLATE 7** Map of the major facilities supporting research in the Arctic.

Everett, 2001; Peterson and Howard-Williams, 2000). Victoria Land contains a number of climatic extremes. Snowfalls vary from almost no precipitation in the McMurdo Dry Valleys to relatively high snowfalls on the northern part of the coast and at Ross Island. Temperature varies from relatively warm temperatures at coastal sites north of the McMurdo Ice Shelf to cold temperatures at the southern inland sites and those adjacent to the Ross Ice Shelf. There is also a great variation in altitude from low to high altitudes along the Trans-Antarctic Mountain range, which stretches the length of the Ross Dependency. Along the Ross Sea coast itself, there are varying degrees of ice cover, including a large polynya, which is a major feature of the Ross Sea. Significant differences in the nonmarine physical environment measurable along the latitudes encompassed by Victoria Land (72-86°S) are found in temperature, solar radiation, humidity, glacier movement, and the biogeochemistry of meltwaters. In the marine environment, differences occur with the dominating influences of light, ocean currents, tides, sea-ice and ice shelf coverage, depth, and the seafloor substrate. Under these conditions, species composition is likely to shift and species diversity may fall as latitude increases with a few remaining species adapted to extreme southern conditions. Ecosystem gradients such as those present along the Victoria Land coast provide a unique natural laboratory for the biochemical, molecular, genetic, physiological, organismal and ecosystem studies that should be considered in future research initiatives.

*Key Questions*

- Are variations in phylotypes across polar latitudes reflected in physiological, molecular, and genetic adaptations to strong environmental gradients?
- Does temperature alone explain latitudinal shifts in biodiversity?
- Can polar latitudinal gradients be used to predict human impacts and global environmental changes?

## POLAR PHYSIOLOGY AND BIOCHEMISTRY

### How Does Living at Extremely Low Temperatures Affect Metabolism and the Cost of Life?

As early as the nineteenth century, natural historians reported that polar marine ectotherms exhibited surprisingly high rates of activity despite their low body temperatures. Terrestrial ectotherms from polar regions were also observed to be highly active at temperatures that would be lethal to tropical or temperate ectotherms. Because rates of metabolic processes like oxygen consumption typically are halved by a 10°C fall in temperature, the question of how polar ectotherms are able to metabolize at high rates at near-freezing temperatures has been an important focus in thermal physiology.

Two closely related issues are of central importance in the analysis of metabolic adaptation to cold. One question focuses on the intrinsic properties of biomolecules (nucleic acids, proteins, membrane lipids, etc.) and asks about the types of adaptations in these molecules that enable them to function satisfactorily at low temperatures where warm-adapted biomolecules are apt to fail. This question is addressed in the next section. Here, a second and closely related question that concerns the rates of metabolic functions in polar species is examined, specifically the extent to which metabolic processes in polar ectotherms are adjusted to compensate for the decelerating effects of reduced temperature.

A number of studies on oxygen consumption by whole organisms (Clarke, 1990; Pörtner, 2002) and measurements of enzymatic activities in individual tissues (Crockett and Sidell, 1990; Kawall et al., 2002) have suggested that rates exhibited by polar ectotherms at their low body temperatures are higher than would be predicted by extrapolating the analogous rates of warm-adapted species down to polar temperatures. This upward adjustment of physiological rates is termed *"metabolic cold adaptation"* or *"temperature compensation of metabolism"* (see Clarke, 1991). Although controversy remains about the extent of cold adaptation and the most appropriate ways to study this phenomenon (Clarke, 1991; Kawall et al., 2002), metabolic compensation to temperature has clearly been achieved in many lineages of polar ectotherms. However, metabolic compensation for polar ectotherms is generally incomplete, such that when rates of oxygen consumption or enzymatic activity are measured at normal body temperatures, the metabolic rates of polar ectotherms are lower than the rates of temperate and tropical species.

One provocative conjecture drawn from these comparative studies is that cold-adapted polar ectotherms can subsist with lower rates of energy turnover than warm-adapted species: The cost of living is reduced in the cold. This conjecture has important implications across the biological sciences spectrum, from ecosystem-level flux of energy among different trophic levels to potential biotechnological approaches to cryopreservation. Although the relationship between temperature and "cost of living" has interested biologists for several decades, the physiological, biochemical, and molecular bases for the apparent abilities of cold-adapted species to reduce their metabolic costs remain largely unknown. Conjectures about the mechanisms underlying the reduced cost of living in the cold include reduced numbers of ion channels, thereby reduced adenosine 5-triphosphate (ATP) demands for transmembrane movement of ions (Hochachka, 1988); reduced costs of protein synthesis in the cold (Marsh et al., 2001); and lowered costs for repair of heat damage to proteins and other biomolecules (Hofmann and Somero, 1995; Somero, in press). There is some evidence supporting each of these proposed mechanisms for reducing the cost of living in the cold.

Recent studies of protein biosynthesis by embryos of an Antarctic sea urchin (*Sterechinus neumayeri*) demonstrated that the rate of protein synthesis is the same in this −1.9 °C Antarctic species as in temperate echinoderms living at much greater temperatures (Marsh et al., 2001). This is an instance where temperature compensation is complete. The basis of the high protein synthetic rate in the Antarctic species appears to be a much higher thermodynamic efficiency of biosynthesis: only 1/25th as much energy is required for protein synthesis in *S. neumayeri* as in temperate species. Because the cost of protein synthesis may represent a large fraction (often approximately 30 percent) of total metabolic activity, reducing these costs while maintaining rates of synthesis could be a critical adaptation for polar species. The mechanistic basis for the high thermodynamic efficiency of protein synthesis in *S. neumayeri* remains to be established. Elucidating these mechanisms might shed light on means for improving the efficiencies of biological processes.

Lowered costs of living may arise not only from low temperatures per se but also from the extreme thermal stability of polar oceans. Stable thermal environments are likely to reduce energy costs because they preclude the necessity of carrying out temperature-acclimatory shifts in the transcriptome and proteome, which are required of ectotherms from thermally variable habitats if their cells are to contain the appropriate types and concentration of proteins and to conduct acclimatory restructuring of lipid-containing structures such as cellular membranes (Hochachka and Somero, 2002). Quantifying the costs of temperature-adaptive changes in the phenotype could contribute important insights to our understanding of organisms' costs of living.

There are at least two reasons why a focus on the cost-of-living issue is pertinent. First, because one potential effect of global warming is an increase in metabolic costs for ectothermic species, it is important to understand more fully how temperature influences these costs, both in terms of basic molecular effects and in the context of how increased energy turnover would affect trophic relationships in polar ecosystems. Ecological effects arising from thermal acceleration of metabolism are apt to be of large magnitude and have a strong impact on species composition. For example, Sanford (1999) showed that increases in water temperature of 3°C led to greatly enhanced predation by sea stars (*Pisaster ochraceus*) on mussels (*Mytilus*). Because the presence of dense beds of *Mytilus* has a strong effect on the structure of rocky intertidal ecosystems, relatively small changes in ambient temperature can be translated into major ecological perturbations. Although Sanford's example comes from the study of a temperate rocky intertidal ecosystem, analogous effects would be predicted for diverse types of polar ecosystems subjected to rising temperatures. Indeed, warming of high-latitude waters is already beginning to influence the distribution limits of aquatic ectotherms. The studies of Welsh and colleagues (1998) on sockeye salmon (*Oncorhynchus nerka*) show that thermal acceleration of metabolic rates of these fish caused by ocean warming may increase their demands for food to levels that cannot be sustained over a large portion of the species' current range. Thus, a truncation of the species' biogeographic distribution toward a more northerly range may be taking place. A further analysis of how temperature changes influence organisms' energy demands and of how the apparently energy-efficient, cold-adapted polar biota would be affected by global warming demands urgent attention.

A second justification for examining how temperature affects metabolic costs concerns biotechnological and biomedical interests. Storage of cells and organs at low temperatures for prolonged periods is important in a number of biomedical contexts, for example, in maintaining banks of materials needed for transplantation surgery. Insights into ways of reducing cellular energy demands might be helpful in designing protocols for improving the longevity and physiological state of cells and organs stored at low temperatures. The strategies used by polar species for reducing energy costs might serve as the basis for new approaches to cryopreservation.

Recently developed genomic, proteomic, and metabolomic techniques offer new avenues for addressing the issue of how temperature affects metabolic processes and the costs of living. Studies of temperature-induced changes in the transcriptome and proteome will provide insights into the extent to which even small changes in temperature influence patterns of gene expression and protein synthesis. Quantification of the effects of changes in temperature on protein biosynthesis and protein degradation (collectively, "protein turnover") will shed light on the manner in which temperature affects one of the most energy-demanding components of physiology. By using a suite of molecular-level approaches to characterize and to quantify temperature's effects on energy budgets for individual organisms, well-grounded predictions about the effects of rising temperatures on complex marine ecosystems can be made, as discussed in the later section, "Physiological and Biochemical Responses to Abiotic Environmental Stresses." Here, then, is a context in which studies using contemporary molecular protocols are poised to make important contributions to global-scale biological issues.

The foregoing discussion of temperature effects on ectothermic species leads to a number of points concerning mammals (endotherms) that are capable of hibernation. As discussed further in Chapter 3, hibernating mammals undergo profound shifts in their physiology during hibernation. Ground squirrels allow the body temperature to plummet to −2.8°C (Barnes, 1989). In contrast to ground squirrels, black bears decrease their body temperatures only by approximately 5°C, yet they too undergo a large suppression of metabolic activity. For both species, metabolic suppression is greater than can be explained strictly on the basis of thermal effects on physiological processes. Other physiological differences distinguish these two hibernators: Ground squirrels lose bone and protein mass but bears do not. Bears maintain perfusion of tissues while in squirrels blood flow may decrease by >90 percent (Boyer and Barnes, 1999).

Understanding the shifts in gene expression that accompany entry into and passage from hibernation will provide important new insights into several issues of biomedical importance, including mechanisms for facultative suppression of metabolism, for sustaining mammalian cells at extremely low temperatures, and for tolerating reduced levels of perfusion.

*Key Questions*

- How does life at low and stable body temperatures affect the metabolic energy requirements (cost of living) of polar organisms?
- What molecular mechanisms allow some polar animals to greatly reduce their metabolic rates?
- How will global warming increase costs of living for polar organisms and what will be the consequence of these effects on polar ecosystems?
- Can the adaptations used by polar organisms to reduce their energetic costs be exploited in biotechnology, for example, in cold preservation (cryopreservation) of cells, tissues, organs, and even whole organisms?
- What mechanisms are used to reversibly suppress metabolism in mammalian hibernators?
- What mechanisms allow hibernating squirrels to withstand large decreases in perfusion of their organs?

## Comparative Analysis of Polar Biomolecules Enhances Understanding of Macromolecular Structure-Function Relationships

Temperature plays a dominant role in the biogeographic distribution of organisms through mechanisms that operate at the molecular, cellular, physiological, and behavioral levels (Hochachka and Somero, 2002).

Given the primacy of temperature, it should not be surprising that some of the most important insights regarding the relationship of molecular structure to function have been established by comparative biochemical analyses of macromolecules from congeneric or confamilial ectotherms of polar and temperate environments. Proteins, both enzymatic and structural, have historically been the favorite objects of these comparative studies. Changes intrinsic to a protein (sequence, three-dimensional shape, posttranslational modification) or extrinsic (cellular milieu) may evolve to alter function.

Enzymes increase reaction rates a millionfold or more by providing alternative reaction pathways from substrate(s) to product(s) that have lower energies of activation. Reaction rates, whether catalyzed or not, are exquisitely sensitive to temperature, doubling or trebling for each increase of 10°C, which at 0°C corresponds to an increase in absolute temperature of only ~4 percent. Thus, polar ectotherms, which in general have evolved from temperate forms, would appear to be at a metabolic disadvantage were they to rely on an unchanged suite of mesophilic enzymes. Yet this is clearly not the case.

Using the glycolytic enzyme lactate dehydrogenase (LDH), Fields and Somero (1998) have shown that the catalytic rate constant $k_{cat}$ of the skeletal muscle isoform of Antarctic notothenioid fishes is four to five times greater than that of mammalian, avian, and thermophilic reptilian orthologues, when each was measured at 0°C. Furthermore, $k_{cat}$ varies in a regular, negative relationship with temperature among fishes, further supporting the hypothesis that the interspecific differences reflect evolutionary adaptation to different thermal regimes. Other enzymes show similar metabolic compensation (Kawall et al., 2002). Since the active-site chemistry of LDH orthologues (and of enzymes in general) is both extremely rapid and evolutionarily conserved, what mechanism or mechanisms could account for such obvious cold adaptation of the enzyme from polar ectotherms? The most plausible explanation is that LDH (and many other enzymes) of polar notothenioids have evolved greater flexibility in regions outside the active site that are responsible for the true rate-limiting step, the attainment of protein conformational changes that enable binding and release of substrate and product. This greater flexibility permits the necessary conformational transitions to occur at lower activation energies. Similar conclusions regarding the importance of flexibility have been drawn from studies of α-amylase from the psychrophilic Antarctic bacterium *Alteromonas haloplanctis* (Aghajari et al., 1998), phosphoglycerate kinase from an Antarctic pseudomonad (Bentahir et al., 2000), the serine protease euphauserase from Antarctic krill (Benjamin et al., 2001), and others; but the molecular strategies that generate the flexibility can be quite variable (for example, decreases in the numbers of disulfide bridges, salt bridges, and/or hydrogen bonds; greater flexibility of surface loops; decreased rigidity of the protein core by reduction of aromatic interactions); (Feller and Gerday, 1997; Gerday et al., 1997; Zecchinon et al., 2001). Further study of enzymes from polar organisms should continue to provide significant new insights into macromolecular structure and function.

*Key Questions*

- Can the proteins of polar organisms teach us general rules about the mechanisms used to alter protein structural stability?
- Are multiple evolutionary mechanisms available to adapt the catalytic power of enzymes?
- What roles do low-molecular-weight organic molecules of polar organisms play in modulating protein stability and function?

**What Are the Biotechnological Applications of Polar Biomolecules?**

The biotechnological potential of enzymes from extremophiles for use in food processing, chemical production, and medical applications has long been recognized (Herbert, 1992). Indeed, one of the early patents awarded to Genentech (Estell, 1988) covered the production of site-directed variants of the protease subtilisin with enhanced activities and stabilities that enable the enzyme to function in detergents across a wide temperature range. Although most attention has been focused on enzymes from thermophiles, including the ubiquitous heat-stable polymerases used in PCR, one can make a compelling case for the energetic advantages of psychrophilic enzymes, which generally do not require heating for activity at mesophilic temperatures (Herbert, 1992; Gerday et al., 2000). Examples of psychrophilic enzymes with biotechnological potential include amylases (Aghajari et al., 1998; D'Amico et al., 2000), chitobiase (Lonhienne et al., 2001), lactases (Hoyoux et al., 2001), DNA ligases (Georlette et al., 2000), lipases (Feller et al., 1991), and proteases (subtilisin, Davail et al., 1994; Narinx et al., 1997; trypsin, Spilliaert and Gudmundsdóttir, 1999).

With the advent of genome-enabled technologies, we can now consider the systematic prospecting of the genomes of polar organisms for other useful enzymes. The application of data-mining strategies to large databases of protein sequences from ectotherms and endotherms may reveal new, general principles that relate protein sequence to thermal performance. This knowledge might, in turn, contribute to the engineering of enzymes with desirable, even novel, functional properties. Finally, the comparative analysis of enzymes from steno- and eurythermal polar organisms will undoubtedly be relevant to predicting the effects of global change on polar communities.

*Key Questions*

- What novel polar bimolecules with practical applications can be found?
- Does cold adaptation of molecules differ in steno- and eurythermal polar organisms, and if so, how?

**Physiological and Biochemical Responses to Abiotic Environmental Stresses**

As alluded to earlier, the general question of how organisms cope with the abiotic (physical and chemical) features of their environment must be considered in the context of two time frames: *long-term evolutionary processes* in which the genetic makeup of an organism is modified, resulting in an increase in "fitness" for a particular niche; and shorter-term *acclimation processes* that enable an organism, over the course of its lifetime, to modify its gene expression to cope with the changing environment. In previous sections, adaptive processes have been discussed. Here, acclimation mechanisms relevant to the polar environment are considered.

Over the course of the year, polar organisms in a variety of niches are subjected to substantial fluctuations in abiotic environmental factors, including temperature, water availability, salinity, and oxygen concentration. Sea ice is an example of one such niche (Thomas and Diekmann, 2002). As surface waters freeze in the autumn, a wide assemblage of organisms become trapped within brine channels that contain the salts expelled from accumulating and coalescing ice crystals. As a result, the organisms rapidly find themselves in a new habitat that presents them with multiple, potentially lethal, abiotic stresses. High salinity is a major factor, as is low temperature that over the winter season extends down to about $-20°C$ (or lower in the Arctic Ocean). In addition, due to biological activity within the confined sea-ice environment, the organisms are subjected to the potentially toxic effects of hypoxia (Gleitz et al., 1995). Then, as the ice melts in response to warming temperatures, the organisms again rapidly find themselves in a dramatically changed environment. Similar fluctuations occur on a larger scale for organisms living on river-impacted Arctic shelves, for example, in the Mackenzie River delta. In winter, the Mackenzie River outflow becomes blocked by what is called a stamukhi zone—ice that ridges not only up but down to the seafloor in many places, as a result of the Arctic ice pack pressing against the coastline—resulting in the formation of "Lake Mackenzie" beneath the ice cover and behind the stamukhi zone. These seasonal ice boundaries thus result in fresh water being trapped on the ocean shelf for a period of months, with full-strength seawater underlying the inverted lake. When the stamukhi zone begins to break up in spring, the fresh water is finally released across the shelf. As a result, significant gradients in salinity exist on the shelf, as a function of time and space, in both the vertical and the horizontal directions. Such salinity gradients are not exclusive to the Mackenzie Shelf, but occur in a pan-Arctic sense across the extensive Russian shelf, where many great rivers flow into the Arctic Ocean. Of course, dramatic fluctuations in physical conditions are not limited to the interface of rivers and oceans. In the Arctic soil environment, for instance, temperatures can range in the surface layers (top 10 cm) from well below freezing in the winter to more than 20°C during sunny periods in summer. On snowmelt, the soil frequently becomes saturated with water and in many cases turns anoxic. As the season advances into summer, soils often then experience severe drying.

Our knowledge in stress physiology is limited. The study of polar organisms can yield some interesting insights on how organisms cope with fluctuating abiotic stresses. For instance, *Colwellia* and other psychrophilic bacteria have been shown to synthesize the ω-3 polyunsaturated fatty acid docosahexaenoic acid (Bowman et al, 1998; Delong et al., 1997). The significance of this is that there is evidence that this and other polyunsaturated fatty acids help organisms maintain membrane homeoviscosity at low temperature. Thus, in the case of *Colwellia* and other polar bacteria, docosahexaenoic acid and other polyunsaturated fatty acids may be critical components of the psychrophilic phenotype. Another significant finding regards dimethylsulfoniopropionate (DMSP), which occurs at high intracellular concentrations in polar and other marine algae. DMSP has been thought to act as an osmolyte, but recent evidence indicates that it and its breakdown products, including dimethylsulfide and dimethylsulfoxide, are also likely to serve as important antioxidants (Sunda et al., 2002). Indeed, DMSP and its breakdown products are very active in scavenging hydroxyl radicals and other reactive oxygen species, and their activities appear to be even greater than those of other well-established antioxidants such as ascorbate and glutathione. Given the important role of DMSP and its breakdown products in tolerance stress, their synthesis increases in response to oxidative stress (Sunda et al., 2002).

Although there is not much information about how polar organisms tolerate abiotic stresses, studies with organisms from temperate environments are yielding important insights into stress tolerance mechanisms that provide a framework for studying polar organisms. For example, there have been recent breakthroughs in understanding the phenomenon of cold acclimation, the process whereby plants and other organisms increase freezing tolerance in response to low, nonfreezing temperatures. For instance, in the model higher plant *Arabidopsis*, a signaling pathway, designated the CBF (CRT-DRE binding factor) cold-response pathway, has been described that has a fundamental role in freezing tolerance (Thomashow, 2001). Within minutes of exposing *Arabidopsis* to low temperature, genes encoding a family of transcriptional activators, known as the CBF (or DREB1) regulatory proteins, are induced, followed by expression of the CBF-targeted genes, designated the CBF regulon (Jaglo-Ottosen et al., 1998; Kasuga et al., 1999; Liu et al., 1998; Stockinger et al., 1997). Expression of the CBF regulon of genes increases the plants' tolerance to freezing through the activation of multiple mechanisms, including the production of proteins (Steponkus et al., 1998) and compatible solutes (Gilmour et al., 2000) that act as cryoprotectants and stabilize membranes and proteins against freeze-induced damage. Expression of the CBF regulon of genes also results in a substantial increase in drought tolerance (Haake et al., 2002; Liu et al., 1998). The cross protection against freezing and drought is largely due to the fact that freezing damage is associated with cellular dehydration. Indeed, many of the cold-inducible genes, including the CBF regulon of genes, are also induced by dehydration stress (Thomashow, 2001). In the case of the CBF regulon, activation occurs through the action of CBF homologues that are themselves rapidly induced in response to dehydration (Haake et al., 2002; Liu et al., 1998). Whether plants from the polar environments have low-temperature pathways related to the CBF pathway and, if so, whether the genes that comprise the CBF regulons are the same as those in *Arabidopsis* or include genes with more potent activities can be addressed through the application of genomic technologies.

    A low-temperature response that is conserved in bacteria is the "cold-shock" response (Weber and Marahiel, 2002; Yamanaka, 1999). When bacteria such as *Escherichia coli* and *Bacillus subtilis* are subjected to a rapid decrease in temperature, they respond rapidly by synthesizing a suite of cold-shock proteins that enable them to acclimate to the change in temperature. Among the cold-induced proteins are those designated cold-shock proteins, abbreviated Csp. These proteins are highly conserved among bacteria and are the most highly expressed in response to cold shock. Current evidence indicates that the Csp have multiple roles in acclimation to low temperature (Weber and Marahiel, 2002; Yamanaka, 1999), including functioning as transcriptional anti-terminators (Bae et al., 2000) and acting as RNA chaperones to facilitate translation of transcripts at low temperature (Jiang et al., 1997). Unlike the heat-shock response, which involves the action of multiple genes that are coordinately regulated by the heat-shock transcription factors, the cold-shock response appears to be controlled by multiple regulatory mechanisms including two-component histidine-kinase systems with membrane-associated environmental sensors that monitor membrane fluidity (Aguilar et al., 2001; Suzuki et al., 2000). Determining whether polar organisms have related regulatory and protective mechanisms or include additional novel mechanisms will be possible through the applications of genomic approaches.

Studies with plants and microorganisms from temperate environments have revealed the existence of multiple mechanisms that contribute to drought and desiccation tolerance. In plants, for instance, genes are induced in response to drought that encode hydrophilic polypeptides known as LEA (late embryo abundant), ERD (early response to dehydration), and COR (cold-regulated), some of which are members of the CBF regulon described above (Thomashow, 2001; Zhu, 2002). These proteins are thought to protect membranes and other cellular structures against dehydration damage (Imai et al., 1996; Steponkus et al., 1998). In plants, bacteria, and other organisms, drought-induced genes also encode enzymes involved in the synthesis of low-molecular-weight compatible solutes such as proline, glycine, betaine, and sugar alcohols that have important roles in both osmotic adjustment and protecting membranes and proteins against damage due to low water potentials (Chen and Murata, 2002; Rontein et al., 2002). Given the extreme nature of the low water availability in the Dry Valleys and many other places in the Arctic and Antarctic, the question arises as to whether the organisms present in these environments have protective mechanisms similar to those described for organisms that inhabit temperate regions or whether they have evolved additional novel mechanisms.

Related to the challenge of low water availability is the challenge of high salinity. This abiotic stress poses not only the problem of low water potential, but also the problems of sodium toxicity and maintenance of ion homeostasis. In plants, high salinity induces the expression of many of the same genes that are induced in response to drought stress, including those encoding LEA and COR proteins and enzymes coding for the synthesis of proline and other molecules involved in osmotic adjustment and mechanisms that prevent dehydration-induced cellular damage (Shinozaki and Yamaguchi-Shinozaki, 2000; Zhu, 2002). In addition, there are mechanisms to remove excess sodium from the cells. For example, in the salt overly sensitive (SOS) pathway, two proteins, SOS3 and SOS2, sense excess sodium and activate a third protein, SOS1, which is a $Na^+$-$H^+$ antiporter that transports sodium out of the cell (Zhu, 2002). The yeast *Saccharomyces cerevisiae* has a similar pathway for ion homeostasis and salt tolerance that includes regulation of sodium and potassium transporters to maintain low $Na^+$ and high $K^+$ concentrations in the cytoplasm

(Serrano et al., 1999). Whether polar organisms have salinity tolerance mechanisms similar to those previously described in nonpolar organisms or have novel mechanisms and determining how these might be affected by the coincident stress of low temperature can be addressed through the application of genomic approaches.

*Key Questions*

Organisms in the Artic and Antarctic are regularly subjected to dramatic fluctuations in abiotic environmental variables such as temperature and salt concentration. Understanding the mechanisms that these organisms have evolved to protect themselves against potentially lethal abiotic stresses is fundamental to our understanding of polar biology. Some of the key questions that can be addressed by genomic technologies are as follows:

- What sensing and regulatory pathways have polar organisms evolved to cope with the dramatic fluctuations in abiotic environmental conditions that occur regularly in the Artic and Antarctic?
- Do polar plants, for instance, have low-temperature pathways related to the CBF pathway; and if so, do they include novel genes or genes with more potent activities?
- Do organisms in sea ice and the McMurdo Dry Valleys have tolerance mechanisms for dehydration and high-salinity stress that are similar to those described in organisms from temperate regions, or have they evolved additional mechanisms?

## POLAR MICROBIAL COMMUNITIES

### How Do Different Microorganisms and Microbial Communities Make Their Living?

One of the greatest current challenges to microbial ecologists is to relate phylogeny and function in complex microbial communities. This challenge is greatest with respect to studies on prokaryotes, particularly heterotrophic bacteria, in which phylogeny (based on rRNA sequence analysis) rarely corresponds to function. In contrast, the function of microbial eukaryotes is better understood. Methods to study the function (or niche) of microorganisms have lagged behind recent advances in elucidation of the phylogenetic diversity and composition of microbial communities (Torsvik and Ovreas, 2002). Thus, even though the new methods have provided substantial information on "who is there," little is known about "what they are doing." Several novel approaches have been developed to relate phylogeny to function, without the need to culture microorganisms. One approach is to incubate environmental samples with $^{13}$C-labeled growth substrates. Subsequently, $^{13}$C-labeled DNA is isolated by density-gradient centrifugation and sequenced (or analyzed by other methods such as electrophoresis) to determine the organism(s) that metabolized the substrate (Radajewski et al., 2000). Alternatively, $^{13}$C-labeled fatty acids are analyzed by isotope ratio mass spectroscopy (Boschker et al., 1998; Bull et al., 2000). Analysis of either the labeled DNA or the fatty acids yields information about the phylogeny of the organisms that incorporated the substrate. A similar approach labels DNA with bromodeoxyuridine (Borneman, 1999; Urbach et al., 1999; Yin et al., 2000). This thymidine analogue is incorporated in the DNA of actively growing cells in a complex community and hence can be identified. The labelled DNA can be stained with fluorescent antibodies for electrophoretic analysis or purified by an immunochemical capture method for subsequent sequencing or other analysis. Phylogeny can also be related to function using a combination of autoradiography and fluorescent in situ hybridization (Cottrell and Kirchman, 2000; Gray et al., 2000; Lee et al., 1999; Ouverney and Fuhrman 2000). In this approach, environmental samples are incubated with $^{14}$C-labeled growth substrates. The whole cells are then hybridized to fluorescently labeled ssu rRNA oligonucleotide probes specific to selected phylogenetic groups. The cells are finally assayed by autoradiography and fluorescence detection, permitting determination of whether particular phylogenetic groups took up the labeled substrate. Such approaches have not been applied to determining metabolic functions of polar microorganisms. Further development of these approaches offers great promise for better understanding functional relationships in complex microbial communities.

Two factors that are consistently implicated as limiting factors for heterotrophic polar soil communities are temperature (Bunnell et al., 1977; Hobbie, 1996; Schimel and Clein, 1996) and water content (Billings et al., 1982; Funk et al., 1994; Robinson et al., 1995;). Soil water content, which can be extreme in polar soils, also influences oxygen availability. Thus, temperature and soil water content largely dictate rates of the critical microbial nutrient cycling activities of organic decomposition and nitrogen mineralization. In her review of this topic, Robinson (2002) concludes that the importance of these two factors is clear, but their interactions with other factors are highly complex. Furthermore, our limited understanding of these complex interactions is a major impediment to predicting the effects of climate change on microbial activities.

*Key Questions*

• What are the functions of the vast range of organisms, most of which are uncultured, in polar microbial communities?
• What are the functional interactions between members of diverse microbial communities?
• What is the reason for, or significance of, the great diversity in most microbial communities?
• How (or are) functional diversity and phyletic diversity related?
• What mechanisms govern microbial metabolic activities that are essential for ecosystem function?
• What are the respective roles of microbial populations and physicochemical factors in controlling microbial nutrient cycling activities?

**How Do the Microbial Components of a Community Interact?**

Microbial and biogeochemical activity in marine waters, including polar seas (Huston and Deming, 2002), occurs in hot spots where microorganisms aggregate on particles (e.g., Azam, 1998). Organisms within these aggregates are thought to operate in a consortial or syntrophic relationship, in which the presence of one type enhances the activity of another (e.g., Murry et al., 1986). An example of a polar microbial consortium (virus-bacteria-cyanobacteria) has been described within the permanent ice covers of the McMurdo Dry Valley lakes (Priscu et al., 1998). A majority of the cyanobacterial and bacterial activity within the ice cover is associated with terrestrially derived sediment aggregates, as opposed to non-aggregated microorganisms embedded in the ice matrix (see Plate 5). On average, lake ice samples that included sediment contain fivefold more bacteria and twofold more viral particles than clear ice sections. Microautoradiographic studies on Dry Valley lake ice revealed that both bacterial and cyanobacterial activities were tightly associated with sediment particles (see Plate 6). Microautoradiographs also indicated that photosynthetically fixed inorganic carbon was a source of organic carbon for heterotrophs. Microzones of low oxygen within the aggregates may be potential sites for $O_2$-sensitive processes such as atmospheric nitrogen fixation (Olson et al., 1998; Paerl and Priscu, 1998). Biogeochemical zonation and diffusional $O_2$ and nutrient concentration gradients likely result from microscale patchiness in microbial metabolic activities (i.e., photosynthesis, respiration). These gradients, in turn, promote metabolic diversity and differential photosynthetic and heterotrophic growth rates. Spatial and temporal relationships within the ice produce a microbial consortium that is of fundamental importance for initiating, maintaining, and optimizing essential life-sustaining production and nutrient-transformation processes (see Plate 6; Paerl and Priscu, 1998; Priscu et al., in press). Close spatial and temporal coupling of metabolite exchange among producers and consumers of organic matter within the ice appears to be the enabling factor that allows microorganisms to coexist in what appears to be an otherwise inhospitable environment. To accomplish this feat, the microorganisms must cooperate in a highly efficient manner. Similar consortia may develop in sea ice where organisms are often concentrated in brine channels. Genomic and proteomic analysis of these communities would reveal the organisms involved and provide important information on the related processes that control their composition and productivity.

*Key Questions*

- What organisms constitute the microbial aggregates?
- What factors control the composition and productivity of the microbial aggregates?
- What attributes do the microorganisms in polar marine ecosystems or permanently ice-covered lakes possess that allow survival in the form of aggregates under harsh conditions?

## Can We Study Polar Microbial Communities as Analogues for the Origin of Extraterrestrial Life?

Today, Earth's biosphere is cold, 14 percent being polar and 90 percent (by volume) cold ocean (<5°C). More than 70 percent of Earth's freshwater occurs as ice, and a large portion of the soil ecosystem exists as permafrost. Indeed, extraterrestrial bodies that have been conjectured to harbor life are icy (Chyba and Phillips, 2001; Wharton et al., 1995). Thus, studies of Earthly ice-bound microorganisms are relevant to the possiblility and persistence of life on extraterrestrial bodies. During the transition from a clement to an inhospitable environment on Mars, liquid water may have progressed from a primarily liquid phase to a solid phase, and the Martian surface would have eventually become ice covered. Martian Orbiter Laser Altimeter images have revealed that water ice exists at the poles of Mars, and subsurface liquid water may be present (Boynton et al., 2002; Malin and Carr, 1999). Furthermore, analyses of Martian meteorites have been used to infer that prokaryotes were once present on the planet (McKay et al., 1996; Thomas-Keptra et al., 2002). Polar ecosystems (see Plates 5 and 6) may serve as models for life on Mars as it cooled (Paerl and Priscu, 1998; Priscu et al., 1998, 1999a,b; Thomas and Dieckmann, 2002), thus assisting the search for extinct or extant life on Mars today (Wharton et al., 1995). Biochemical traces of life or even viable microorganisms may be well protected from destruction if deposited within polar perennial ice or frozen below the planet's surface. During high obliquity, increases in the temperature and atmospheric pressure at the northern pole of Mars (Malin and Carr, 1999; McKay and Stoker, 1989) could result in a discharge of liquid water that might create environments with ecological niches similar to those inhabited by microorganisms in terrestrial polar and glacial regions. Periodic effluxes of hydrothermal heat to the surface could move microorganisms from the Martian subterranean, where conditions may be more favorable for extant life (McKay, 2001). Annual partial melting of the ice caps might then provide conditions compatible with active life or at least provide water in which these microorganisms may be preserved by subsequent freezing (Clifford et al., 2000; McKay and Stoker, 1989). We can evaluate such hypotheses by analysis of polar ecosystems, but assessment of their validity will depend, ultimately, on scientific missions to explore and study the frozen surface of Mars.

Surface ice on Europa, one of the moons of Jupiter, appears to be in contact with subsurface liquid water (Kivelson et al., 2000). Geothermal heating and the tidal forces generated by orbiting Jupiter are thought to maintain a 50-100-km-deep liquid ocean on Europa with perhaps twice the volume of Earth's ocean (Chyba and Phillips, 2001) but beneath an ice shell at least 3-4 km thick (Turtle and Pierazzo, 2001). Cold temperatures (<128 K) (Orton et al., 1996), combined with intense levels of radiation, would appear to preclude the existence of life on the surface. Moreover, the zone of habitability (where liquid water is stable) may only be present kilometers below the surface, where sunlight is unable to penetrate (Chyba and Hand, 2001). Europa's surface appears strikingly similar to terrestrial polar ice floes, suggesting that the outer shell of ice is periodically exchanged with the underlying ocean. The ridges in the crust and the apparent rafting of dislocated pieces imply that subterranean liquid water flows up through stress-induced tidal cracks, which may then offer provisional habitats at shallow depth for photosynthesis or other forms of metabolism (Gaidos and Nimmo, 2000). Gaidos et al. (1999) argue that without a source of oxidants, Europa's subsurface ocean would be destined to reach chemical equilibrium, making biologically dependent redox reactions thermodynamically impossible. However, the surface is continually bombarded with high-energy particles, producing molecular oxygen and peroxides, as well as formaldehyde and other organic carbon sources (Chyba, 2000; Chyba and Hand, 2001), and Europan microbial life may conceivably subsist without employing photosynthetic or chemoautotrophic life-styles. In this scenario, mixing between the crust and subsurface need not be the only mechanism required to supply organics and oxygen at levels sufficient to support life (Chyba, 2000). Tidal heat generation and electrolysis might also provide sources of energy that could be coupled to bioenergetic redox reactions (Greenberg et al., 2000). The vast network of Antarctic subglacial lakes that lie ~4 km beneath the permanent ice sheet provide an Earthly analogue for life on Europa and may serve as a model system to develop the noncontaminating technologies that will be required to sample Europa. Wintertime Arctic sea ice, where liquid brines exist at temperatures of –35°C, provides a marine model for exploring the limits of life on this planet of relevance to Europa's saline, ice-covered ocean (Deming, 2002).

*Key Questions*

- How does the combination of high pressure and low temperature in deep Antarctic lakes or the deep Arctic basins influence microbial survival and activity?
- What are the tolerance levels of various stressors (for example, salinity, radiation, and heavy metals) at the lower temperature limit for microbial life?
- What is the lower temperature limit for evolving life?
- Can molecular probes developed from organisms living in Earth's polar environments be used in future extraterrestrial life detection?

## What Are the Important Biogeochemical Processes That Have to Be Measured?

Genome-enabled techniques have played key roles in advancing our understanding of important geochemical processes. They have allowed us to assess the diversity of organisms involved in specific processes, to study the distribution of key organisms, and to evaluate some of the factors that regulate expression of these pathways. Key breakthroughs in this area include the discovery of a completely unsuspected mode of phototrophy in marine bacteria (Béjà et al., 2000, 2001), the demonstration that nitrogen fixation is widespread in the open ocean (Zehr et al., 1998, 2001), and the demonstration that a *Nitrosospira*-like bacterium is widely distributed in the ocean and that it, rather than *Nitrosococcus*, may be responsible for much of the ammonium oxidation in the open ocean (Figures 2-4 and 2-5); (Bano and Hollibaugh, 2000; Hollibaugh et al., 2002).

Key geochemical processes of relevance to polar biology that can be studied today using genome-enabled techniques include photosynthesis, nitrogen fixation, nitrification and denitrification, sulfate reduction, methanogenesis and methane oxidation, and metal (Fe, Mn) or metalloid (Se, As) redox reactions. Two major challenges in this area of research are (1) connecting a functional gene detected by the use of a specific set of

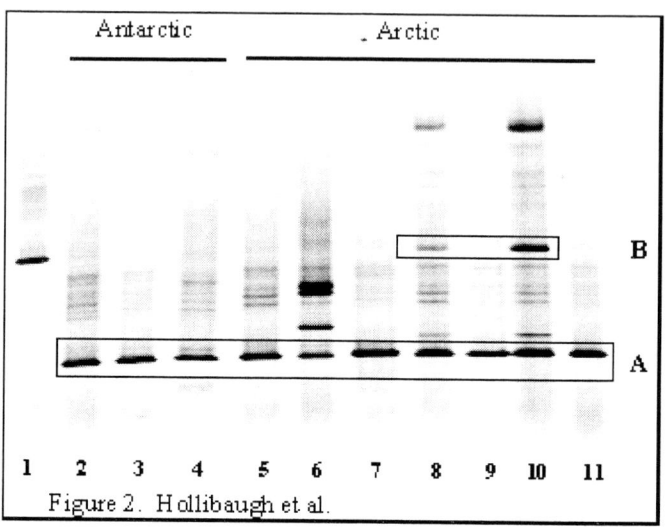

**FIGURE 2-4** Comparison of β-proteobacterial ammonia-oxidizing bacteria (AOB) in polar oceans. PCR was used to amplify portions of ssu ribosomal genes from AOB harvested from water samples. The products of the mixed template amplification were resolved by denaturing gradient gel electrophoresis. The band contained in Box A is from a *Nitrosospira*-like AOB found in both oceans, while the band contained in Box B is from a novel *Nitrososmonas*-like AOB that was found only in Arctic Ocean samples. SOURCE: Hollibaugh et al., 2002.

PCR primers to the rest of the genome and, thus, to the additional genetic capability; and (2) developing robust probes and primer sets that can be used to detect poorly conserved genes encoding proteins that participate in important biogeochemical reactions.

Some progress has been made in the connection of a functional gene with the rest of the genome by the application of large insert cloning vectors (bacterial artificial chromosomes or fosmids, [Béjà et al., 2000; Stein et al., 1996]); however, this approach is laborious and depends on the isolation of large quantities of high-molecular-weight DNA. Furthermore, it is a shotgun approach, so organisms are represented in the library in approximately the same relative abundance as in the original sample. This means that rare organisms are not likely to be detected by this approach and there is currently no way of enriching for selected genomes. Often, as in the case of the *Nitrosospira*-like organism mentioned above, organisms of biogeochemical interest are relatively rare and thus will not

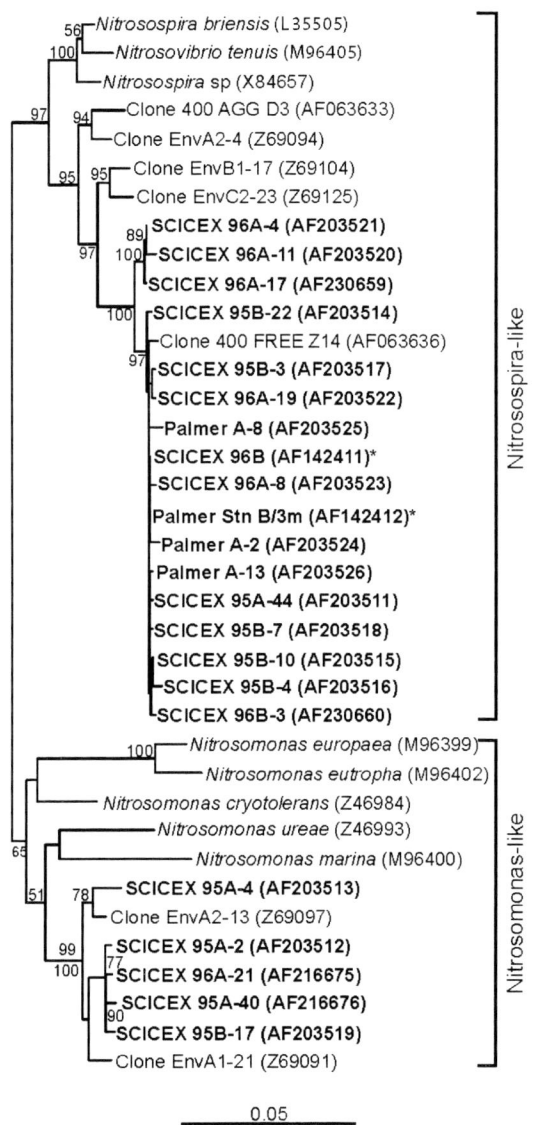

FIGURE 2-5 Phylogenetic tree constructed from sequences of cloned ssu rRNA genes showing the relationship of Arctic and Southern Ocean β-proteobacterial ammonia-oxidizing bacteria (AOB) sequences to cultured representatives of the AOB. Arctic clones are indicated by the prefix "SCICEX," Antarctic clones are indicated by the prefix "Palmer." Arctic and Antarctic *Nitrosospira*-like sequences are essentially identical over 1,040 bp and cluster together. This organism does not have any close relatives in culture. SOURCE: Hollibaugh et al., 2002.

be sampled by this approach. At present, this is a stumbling block with no obvious solution other than to use selective growth conditions to enrich, or isolate into pure culture, the organism of interest. However, the apparently low cultivability of many bacteria constrains this approach as tightly as relative abundance constrains information provided by shotgun cloning.

At this point, only genes for which enough sequence information is available to construct primers or probes can be detected. Although the physiology of some processes seems to be relatively conserved (for example, ammonia oxidation) so that phylogenies based on ssu rRNA genes and on functional genes of the pathways are congruent (Purkhold et al., 2000), others are not (for example, denitrification, oxidation of organic compounds, nitrogen fixation or metalloid reductases) (Niggemyer et al., 2001). While it is not necessarily true that functional genes for processes that are polyphyletic with regard to ssu rRNA genes will be similarly diverse, this has been demonstrated for some genes (for example, arsenate reductase) (Stolz and Oremland, 1999). Such variability complicates primer design.

The discovery of unsuspected biogeochemical pathways (for example, anaerobic methane oxidation) (Boetius et al., 2000; Michaelis et al., 2002) and the demonstration that microorganisms participate in reactions that were thought previously to be abiotic (Oremland et al., 2002) further complicates biogeochemical analysis. Distribution of these phenotypes in microbial communities is presently unknown; thus, there may be additional types of bacteria or archaea that mediate these newly discovered biogeochemical reactions. Detection of additional types of bacteria and archaea may prove difficult without a search image. In the context of polar biogeochemistry, undersampled habitats that might lead to the discovery of novel organisms are deep Arctic waters, nepholoid (particle-rich) waters, Dry Valley lakes and soils, subglacial lakes, ice cores, and other polar soils.

A major impediment to the study of biogeochemical processes is the inefficiency with which bacteria can be cultured. Most traditional culturing approaches yield a small proportion of the bacteria present in a given environment. Connon and Giovannoni (2002) and Rappé et al. (2002) recently developed and applied successfully a high-throughput method for culturing (HTC) previously unculturable prokaryotes that thrive in dilute environments (oligotrophs). Although the HTC system is similar in some respects to microtiter dish screens; it is designed specifically for detection, phylogenetic identification, and isolation of organisms that can only achieve low cell densities in laboratory culture. HTC addresses at least four problems that likely make conventional culturing of oligotrophic prokaryotes impossible: (1) For as yet unknown reasons, these organisms may be able to grow only to low cell densities; (2) growth may occur only within narrowly defined culture conditions; (3) growth may require a second (or more) organism; and (4) growth may be inhibited by contaminants in laboratory reagents. HTC uses extinction culturing to propagate organisms at substrate concentrations and cell densities that are typical of natural waters but significantly lower than those of laboratory media used for conventional culturing. HTC can detect and identify cells after fewer than 12 cell divisions, which shortens considerably the time required for experiments with slow-growing cells (Connon and Giovannoni, 2002). Thus, HTC is ideally suited for work with psychrophiles, which often have slow growth rates. Recently, the efficacy of HTC was validated by its use for the isolation of members of the ubiquitous SAR11 marine bacterioplankton clade, organisms of global significance that previously had eluded cultivation (Figure 2-5; Rappé et al., 2002).

*Key Questions*

- Is there a relationship between composition and biogeochemical function in polar microbial communities? Can we infer rates of processes from genomic data?
- What factors control the expression of various biogeochemically significant pathways?
- How much functional redundancy is there in polar microbial assemblages?
- What is the relationship between microorganisms facilitating important biogeochemical reactions in polar environments and those performing the same function at lower latitudes?
- How does the effect of temperature on rates of microbial processes influence the net biogeochemical performance of polar oceans?
- Why do polar microbial processes have larger temperature coefficients than the same process at lower latitudes?

## HUMAN IMPACTS

*"My aunt, Mabel Toolie, said [to me]: "The Earth is faster now." She was not meaning that the time is moving fast these days or that the events are going faster. But she was talking about how all this weather is changing. . . .* (Pungowiyi, 2000, in Krupnik and Jolly, 2002)

Pungowiyi's oral report provides a compelling reminder that the environment of the polar regions is changing. Some of this change is anthropogenic in origin. The indigenous human populations of the north not only are subject to this change but also are active participants in studying these phenomena. Here, the committee examines some important human impacts on polar biota and addresses how genomic technologies can be applied to understand them.

## The Arctic and Antarctic Ozone Holes: Impacts of Elevated Ultraviolet Irradiance on Polar Biota

The concentration of stratospheric ozone has decreased significantly during the past three decades, the result of catalytic destruction mediated by the photodegradation products of anthropogenic chlorofluorocarbons (Anderson et al., 1991; Schoeberl and Hartmann, 1991). Ozone depletion has been most dramatic at the poles (Frederick et al., 1998; Hofmann, 1996), especially over Antarctica where ozone levels typically decline >50 percent during the austral spring "ozone hole" (Frederick and Snell, 1988; Solomon, 1990), and further depletion over a broader geographical range is anticipated over the next 25-100 years (Crawford, 1987; Jones and Shanklin, 1995). Atmospheric ozone strongly and selectively absorbs solar UV-B (280-320 nm), thus reducing the intensity of the most biologically damaging solar wavelengths that penetrate the atmosphere (Molina and Molina, 1986), and decreased stratospheric ozone has been linked directly to increased UV-B flux at Earth's surface (Lubin et al., 1989). UV-B also penetrates to ecologically significant depths (20-30 m) in the ocean at intensities that can cause measurable biological damage (Calkins and Thordardottir, 1980; Jeffrey et al., 1996; Karentz et al., 1991; Smith and Baker, 1979; Smith et al., 1992). Therefore, the fitness of polar, especially Antarctic, terrestrial marine organisms in coastal regions and the upper photic zone of open oceans may be affected deleteriously by the projected long-term increase in UV-B flux (Cullen and Lesser, 1991; Jeffrey et al., 1996).

The impact of elevated UV-B has been documented most extensively for the primary producers of polar marine ecosystems (de Mora et al., 2000; Neale et al., 1998; Smith et al., 1992; Prézelin et al., 1994; Smith et al., 1994; Weiler and Penhale, 1994). Primary productivity in the Southern Ocean declines by as much as 15 percent in areas affected by the ozone hole (Smith et al., 1992), but vertical mixing of the water column can mitigate the decrease (Neale et al., 1998). Recently developed models of primary production incorporate multiple linked variables, including the interactive effects of UV and visible radiation (Neale et al., 1998; Prézelin et al., 1998) and the mechanisms of DNA repair in phytoplankton (Neale et al., 2001).

Two other factors that should be addressed in modeling UV effects are nutrient limitation and temperature. Litchman et al. (2002) have generated biological weighting functions that quantify the effect of UV on dinoflagellate cultures grown under nitrogen-limited and nutrient-replete conditions. They found that nutrient-limited cultures are 1.5 times more sensitive to UV than nutrient-replete cultures. Furthermore, UV exposure inactivates nitrogen metabolism and affects both nitrate and ammonia uptake (Vernet, 2001, and references cited therein). Thus, UV and nutrient limitation may have a compound effect on productivity. Although polar oceans are generally thought to be plentiful in macronutrients such as nitrogen and phosphorus, some regions of the Southern Ocean are seasonally iron-limited (for example, the Ross Sea) (de Baar et al., 1995; Olson et al., 2000). The combined impact of nutrient limitation and UV on polar marine organisms at constant cold temperature is an important subject that is readily amenable to analysis using genomic technologies (for example, transcriptome profiling using microarrays).

Cold temperatures may exacerbate the negative impact of UV on populations of polar organisms because the enzymatic systems that repair UV-mediated DNA damage (Cullen and Lesser, 1991; Lesser et al., 1994) are temperature sensitive (Pakker et al., 2000). However, Ivanov et al. (2000) have shown that cultures of a filamentous cyanobacterium, *Plectonema boryanum*, grown under low temperature are, in fact, more resistant to acute UV exposure than those grown at moderate temperature, in part because cold induces the accumulation of photoprotective (UV-absorbing) pigments (for example, carotenoids, scytonemin, and mycosporine-like amino acids). To account for the great differences in UV protection and DNA repair rates by natural assemblages of Antarctic phytoplankton (Neale et al., 2001), it is critical to investigate regulation of the protection and repair systems of these organisms at the level of their transcriptomes and proteomes.

Although most studies on Arctic and Antarctic marine phytoplankton address the effect of UV-B on overall productivity of a community, increasing UV-B fluxes may favor phytoplankton species that are resistant to UV-B, thus leading to shifts in community composition. Karentz et al. (1991) reported that 12 species of Antarctic diatoms varied widely in their molecular and cellular responses to UV exposure, such that smaller cells with higher surface-to-volume ratios sustained more damage per unit DNA. These results imply that increased UV-B fluxes due to ozone depletion may influence the size and taxonomic structure of phytoplankton. Community shifts of phytoplankton as a result of ozone depletion have been observed in incubation experiments (Davidson et al., 1996), but it is unknown whether such shifts occur in natural assemblages. Because evaluation of the photophysiology of single phytoplankton cells is now feasible through the use of microspectrofluorometers, assessment of the UV sensitivity of individual species in a natural assemblage has become possible and may lead to the ability to predict phytoplankton community changes. However, a thorough evaluation of phytoplankton sensitivity to UV exposure must be based on identification of the multiple, UV-sensitive targets (nucleic acids, proteins, lipids) that affect photosynthesis, growth, and reproduction (Vincent and Neale, 2000).

The cascade of trophic events that result from UV-B perturbation of phytoplankton community structure is largely unexplored and may have important ramifications for zooplankton, fish, and mammalian populations. Linkage of whole ecosystem studies to measurements of the molecular responses of individual species will be critical to understanding the trophic impacts of UV radiation (Mostajir et al., 2000, and references cited therein) and validating predictive ecosystem models (Day and Neale, 2002). Furthermore, long-term monitoring of natural community ensembles will be necessary so that changes induced by environmental stressors such as UV radiation can be differentiated from natural background variability.

The decrease in bacterial production caused by UV radiation is comparable to the decrease in phytoplankton production in percent inhibition under similar conditions (Jeffrey et al., 2000). Because of their small size, bacterioplankton are almost exclusively dependent on repair mechanisms to counteract UV effects. Most DNA damage in Antarctic plankton is thus associated with bacteria (Buma et al., 2001; Meador et al., 2002). Bacterioplankton community sensitivity to UV radiation appears to be related to ambient solar irradiance. Near Palmer Station, bacterioplankton were observed to display decreased UV sensitivity as day length and solar irradiance increased from early spring through summer (Jeffrey et al., 2000; Pakulski et al., in preparation). Similarly, spatial variability in bacterioplankton sensitivity to UV along a latitudinal transect was related to incident solar irradiance. Samples collected from lower latitudes were observed to be less sensitive to UV radiation than those collected in low-light/high-latitude environments (Pakulski et al., in preparation). Whether such differential sensitivity of bacterioplankton populations is due to acclimatory adjustments by the community as a whole or to the selection of resistant bacteria species is an important question that can be approached using genomic techniques.

Finally, increased ultraviolet irradiance has been shown to impact metazoan planktonic groups, including zooplankton (e.g., krill), fish eggs (Malloy et al., 1997), and the eggs and larvae of benthic invertebrates (Karentz and Bosch, 2001). These effects appear to be related primarily to UV radiation-induced DNA damage, thus emphasizing the need to understand the molecular mechanisms and regulation of DNA repair systems in Antarctic organisms and the role of repair in modulating UV stress.

*Key Questions*

• How do UV stress, nutrient status, and temperature changes interact to influence microbial productivity?
• How does low temperature affect the regulation of UV acclimation or UV repair mechanisms in polar marine organisms?
• Do differential effects of UV stress on different classes of organisms lead to an ecosystem shift?

## Introduced Species and Diseases: Genomic Monitoring and Impact Assessment

The potential for major ecological impacts of introduced species in polar environments is an important concern. This concern is greatest in Antarctica where humans were not present until the very recent past. Regional warming and increased human visitation in Antarctica are increasing the likelihood of introductions of exotic species with unknown impacts on polar biodiversity and ecosystem functioning. A second major area of concern is infection by human-introduced diseases of wildlife having little or no natural resistance to foreign pathogens. Third, there is an urgent need to monitor fish stocks and to prevent the exploitation of commercial fisheries and the illegal harvesting of protected species.

### Vascular Plants

Human visitation is increasing dramatically in Antarctica, and regional warming along the Antarctic Peninsula and sub-Arctic islands is increasing the likelihood of introductions of exotic species (Bergstrom and Chown, 1999). To the committee's knowledge, there have been no recent plant invasions in Antarctica. Either past introductions of vascular plants along the Antarctic Peninsula (a practice now banned by the Antarctic Conservation Act), have failed over time, or the exotics have been destroyed (Lewis-Smith, 1996). As the potential for introductions grows, genomic approaches will provide the tools necessary to trace the sources and the spread of invasive species in polar regions.

*Key Questions*

• How great are the risks from human-introduced species to polar ecosystems?
• What genetic factors predispose an organism to being a successful invader?

- How do invader species influence the composition of the polar foodwebs and functioning of ecosystems?

**Monitoring the Introduction of Diseases into Polar Regions**

Since 1987, morbillivirus (MV) infections of aquatic mammals, chiefly pinnipeds and cetaceans, have caused serious disease outbreaks and high mortality (Visser et al., 1993). Subsequent research has revealed that three MV species of the Paramyxoviridae family are responsible for these epidemics (Saliki et al., 2002): canine distemper virus in seals and polar bears, cetacean morbillivirus in dolphins and porpoises, and phocine distemper virus in pinnipeds. Although such outbreaks have not been documented conclusively in polar regions, evidence of prior canine-distemper-like MV infections has been found in Arctic and Antarctic seal populations (Have et al., 1991); and mass die-offs of Antarctic seals and penguins may have been due to human-introduced infections (Kerry et al., 2002). Concerns about the potential effect of foreign disease prompted measures in the Antarctic Treaty (for example, a ban on dogs) to prevent such incidents (<http://eelink.net/~asilwildlife/antarctic1964.html>). There is an urgent need to use genomic technologies to differentiate between natural and introduced MV outbreaks among polar mammals so that we can better understand the risks to these populations.

*Key Question*

- How can we best deploy genomic methods to trace, and ultimately remediate, the effects of introduced diseases?

**Native and Farmed Salmon**

Salmon have played a major cultural, nutritional, and economic role in North Pacific communities for thousands of years. Currently, major threats are facing salmon populations, and contemporary genomic methods are likely to prove useful for weighing these threats and predicting the future success or failure of salmon stocks.

As discussed earlier in the section "How Does Living at Extremely Low Temperatures Affect Metabolism and the Cost of Life," salmon populations are seriously threatened by warming of the oceans (Welsh et al., 1998). As surface temperatures rise, sockeye salmon will be increasingly excluded from lower-latitude waters. If present warming trends persist throughout this century, Welsh et al. (1998) predict that sockeye salmon could be excluded from the Pacific Ocean. The economic, cultural, and ecological consequences of this change would be severe. Therefore, increased study of oceanic populations of salmon to further evaluate the extent of this threat is critical. To date, most experimental work with salmon has involved freshwater stages of these species; work on oceanic populations merits a high priority. Studies of oceanic stages of salmon should involve remote-sensing efforts, using state-of-the-art microprocessor technologies, to establish a strong linkage between water temperature and distribution. These data could then be integrated with physiological and genomic data to evaluate the capability of salmon to acclimate or adapt to changing ocean conditions.

Genetic characterization of salmon populations is warranted for several reasons. Genetic techniques could test the conjecture of Welsh et al. (1998) that all populations of salmon are similarly threatened by warming. If all populations are being similarly affected by warming, there may be no reservoir of less heat-sensitive salmon available to replenish losses incurred by heat-sensitive stocks. Genetic methods come into play in another key arena: monitoring the entry of wild populations of genes into farmed salmon. Studying the potential entry of farmed fish, including genetically engineered animals (Masri et al., 2002), into natural ecosystems is urgent in view of expansions in fish-farming operations. Farmed fish may also facilitate the dispersal of pathogens into wild populations, especially as brood stocks are moved around the world. These introduced pathogens can be identified using genomic techniques.

*Key Question*

- How will global climate change affect the biogeography of polar organisms?

## Genomic Technologies to Monitor Stocks of Commercially Exploited Marine Organisms

Genomic tools also have roles in assessing the impacts of humans on the world's fisheries. One application addresses the adherence of nations to international agreements. Through the use of "molecular forensics," it has been established that the terms of multiple whaling moratoria have been flagrantly violated. For instance, six baleen whale species and the sperm whale have been protected by international agreements dating from 1989 or earlier. Yet the use of molecular markers provided evidence that eight species of baleen whales and sperm whale products were among those purchased in Japanese markets from 1993 to 1999 (Baker et al., 2000a,b). Overall, protected species accounted for about 10 percent of the whale products from these markets. Indeed, genomic tools can be incredibly precise, enabling Cipriano and Palumbi (1999) to trace the life of an individual protected whale from its conception in the North Atlantic in 1964 to its sale as raw meat in Osaka, Japan, in 1993. Genomic approaches can also be used to differentiate fish species subject to legal exploitation from those that are pirated illegally. The two congeners of *Dissostichus* merit study in this regard, because the strong market for the South American species *D. eleginoides* may be being satisfied in part by illegal fishing of the Antarctic congener *D. mawsoni*.

*Key Question*

• How can genomic tools or data be used to monitor introduced species and illegal harvesting of protected species?

## SUMMARY

This chapter has described examples of the compelling opportunities for intensified research on polar ecosystems and the major advantages that would accrue through application of genome sciences and other enabling technologies to these problems. This chapter is by no means an exhaustive review of all the exciting research in polar biology that can benefit from genomic technologies and may reflect the expertise on the committee. Examples of other issues in polar biology that may be addressed by genomic studies include:

• *Biological rhythms, ultradian, circadian, and circannual cycles of resident plants and animals in the Arctic and Antarctic.* These regulatory systems—and their space persistence, mechanisms of entrainment, and physiological and behavioral functions—are largely unstudied in polar organisms. The investigation of biological rhythms in organisms subjected to extreme light/dark cycles may provide insights into the genetic and molecular structure and function of biological clocks of all organisms.
• *Molecular and endocrine mechanisms underlying migration and reproduction in breeding birds in polar regions.* The genetics of migration and orientation can help assess the impact of climate change on migratory birds in polar regions. The regulation of migratory timing and distances may or may not be flexible enough between generations to allow individuals or populations or species to respond to rapid changes in climate at their breeding grounds (Both and Visser, 2001).
• *Biodiversity of organisms in hot vents in the Arctic Ocean.* Vents at lower latitudes release hyperthermophiles into ambient waters at 2-4°C. In the Arctic, they would be released into −1.7°C waters. If these colder waters preserve the hyperthermophiles more efficiently, novel strains expressing novel DNA polymerases and other enzymes of interest may

be discovered. The invertebrate colonizers of the ambient waters surrounding Arctic vents are likely to differ from those in lower-latitude waters. The genomics of these geographically isolated micro- and macro-organisms should provide important new information on the evolution of life at extreme temperatures, both hot and cold.

• *Episodic food supply.* In Arctic marine waters, researchers are studying the response of benthic animals and microorganisms to the early seasonal pulse of ice algae to the seafloor. Currently, the distinction between the ice algae that arrived early at the seafloor from phytodetritus (from phytoplankton) and those that arrived later in the season is difficult. Genomic markers that distinguish ice algae from phytoplankton would be invaluable to this line of research.

• *Snow ice as a habitat for microorganisms.* Snow ice is one of the habitats (along with sea ice and permafrost) recently shown to support microbial activity at a temperature extreme of $-17°C$ (Carpenter et al., 2000). Genomic work on the responsible microorganisms in snow (as in any form of ice) would increase our knowledge of the lower temperature limit of life and what constrains it.

Although some of the research questions in polar biology put forward in this chapter apply to temperate and tropical regions, pursuit of these studies in the polar ecosystems cannot be neglected because (1) the polar regions are one of the least studied and understood ecosystems; (2) genome research applied to polar biology would serve as a useful "test bed" for temperate and tropical regions (e.g., there are tens of thousands of tropical fishes but only about 250 in Antarctica); and (3) comparative studies across latitudinal clines can elucidate physiological and biochemical mechanisms for adaptation. Subsequent chapters outline a strategy for implementation of this increasingly important research agenda.

# 3

# The Polar Genome Science Initiative

Evolutionary processes have created many biological communities that are as stunning in their beauty and complexity as they are unexpected by and novel to biological scientists (Diamond, 2001). Some of the most fascinating and diverse of these natural experiments have occurred among the organisms and biological communities of the polar regions. Effective strategies for exploring polar ecosystems using approaches based on genome science and other technologies can rapidly advance our understanding of these ecosystems.

## SELECTION OF ORGANISMS AND CONSORTIA FOR GENOME ANALYSIS

The success of the publicly and privately financed human genome initiatives (Lander et al., 2001; Venter et al., 2001) is directly attributable to the development of high-throughput, low-cost DNA sequencing technologies and appropriate bioinformatics tools for assembling and annotating the approximately 3 billion base pairs (bp) of the human genome. Clearly, the sequencing of genomes should no longer be constrained to "model" organisms or limited by resource considerations, and the Polar Genome Science Initiative need not focus on technology development. Nevertheless, the selection of organisms or consortia must be guided by appropriate criteria. The committee proposes that selection of an organism or consortium be based on evidence that:

- analysis of its genome will address broad and significant scientific questions;
- it is a good model for evolution in an isolated polar environment;
- it provides opportunities for comparisons with organisms of comparable ecotype from polar habitats and along polar-to-temperate latitudinal clines; or
- its cellular processes possess characteristics of biotechnological or clinical interest.

Based on these criteria, the committee provides examples of polar species and consortia that fit the selection criteria mentioned above, but certainly other organisms may fit the selection criteria and warrant sequencing in the near term. The knowledge gained from these organisms will provide an invaluable framework for identifying other organisms for future sequencing projects. Whether some or all of the organisms listed below or other polar organisms are selected for genome analysis will depend on the availability of funding and on changes in research priorities.

### Prokaryotes

Efforts in prokaryotic microbial genomics over the past decade have provided a wealth of information on the nature of microbial diversity and the forces that shape prokaryotic genomes. To date, more than 80 prokaryotic genomes have been sequenced completely (<http://www.tigr.org/tigr-scripts/CMR2/CMRGenomes.spl>), with many more in progress. This collection of data contains more than 250,000 bacterial genes from phylogenetically diverse species; however, it likely represents less than 0.1 percent of the globally distributed prokaryotic gene pool (Stahl and Tiedje, 2002). Remarkably, however, no genome sequence has yet been completed for a psychrophilic prokaryote; and to the committee's knowledge, only two are in progress (isolates of *Colwellia* and *Psychrobacter*). This is a shortcoming that should be remedied since psychrophilic prokaryotes offer a tremendous opportunity to better understand the genetic basis for the psychrophilic phenotype, which has potential biotechnological applications as outlined in Chapter 2.

Because of their small genomes, it is completely within reason to obtain near full-genome sequences for numerous (20 or more) prokaryotes isolated from Arctic and Antarctic locations including sea and freshwater ice, permafrost, and other extremely cold niches. By comparing the genetic "informational content" of a large number of psychrophilic prokaryotes with each other and with the genetic complements of psychrotolerant and mesophilic prokaryotes, especially those belonging to the same or closely related genera, it should become apparent whether the psychrophilic phenotype is associated with the presence of specific genes that are not found in nonpsychrophiles, whether psychrophiles have conserved themes in gene complements that distinguish them from their nonpsychrophilic counterparts (for example, encoding enzymes for the synthesis of specific lipids or fatty acid derivatives well suited for life at low temperature, the synthesis of osmolytes that have cryoprotective properties), or whether there are genome features (for example, number of duplicated genes, abundance of mobile genetic elements) that might play a role in adaptation to the cold. In addition, the sequence information would serve as the basis for conducting functional genomic studies (transcriptome, proteome, and metabalome analysis) to determine, for instance, whether growth at low temperature involves differences in the abilities of psychrophiles and nonpsychrophiles to express their genome complements at low temperature (for example, the synthesis of transcripts and polypeptides) or whether it relates to the activities of certain proteins at low temperature.

Within the criteria outlined at the beginning of this chapter, top candidates for sequencing would be representative psychrophilic bacteria, particularly organisms that have closely related psychrotolerant isolates for comparison. One project currently in progress involves sequencing the genome of a representative *Colwellia* sp., which belongs to the gamma-Proteobacteria. The *Colwellia* being sequenced is an obligate psychrophile isolated from Arctic sediments. It would be prudent to include psychrophiles from other phylogenetic groups for comparison. Other representatives of the bacterioplankton community might include polar representatives of the SAR 11 group, which are abundant in polar bacterioplankton communities (Bano and Hollibaugh, 2002). These prokaryotes may or may not be psychrophiles. A temperate representative of this group has just been cultured and is currently being sequenced. Given that SAR 11 small subunit (ssu) ribosomal ribonucleic acid (rRNA) gene sequences from polar and temperate environments are slightly but consistently different (Bano and Hollibaugh, 2002; Martinez and Valera, 2000), it is likely that polar populations differ from temperate or tropical representatives of this group. Other groups of polar prokaryotes that should be considered for sequencing include those isolated from the Siberian permafrost (Vishnivetskaya et al., 2000; Vorobyova et al., 1997) and the low-temperature Crenarchaeota that have been shown to dominate Antarctic plankton communities at times (Delong et al., 1994; Massana et al., 1998; Murray et al., 1998). Unfortunately, the latter group of organisms does not yet have any representatives in culture. As further research unravels the ecology and physiology of polar plankton communities, other candidates for genome sequencing will become obvious.

*Cyanobacteria.* Cyanobacterial mats dominated by oscillatorians are a feature of streams, lakes, and ponds in both Arctic and Antarctic regions (Vincent and Neale, 2000; Priscu et al., in press) and constitute a major component of autotrophic community biomass and productivity in these polar deserts (Priscu et al., 1998; Vézina and Vincent, 1997; Vincent et al., 1993). Surprisingly, although they are abundant in temperate and tropical oceans, marine cyanobacteria have not been found in polar waters. Although the polar freshwater ecosystems are predominantly cold, with summer temperatures rarely exceeding 0°C, most cyanobacteria isolated from these habitats are psychrotolerant and show optimal growth and photosynthesis at 15°C or higher (Fritsen and Priscu, 1998; Tang et al., 1997a). These data imply that polar cyanobacteria evolved from temperate latitudes and later colonized polar regions (Seaburg et al., 1981; Vincent and James, 1996). Recently, Nadeau and Castenholz (2000) described the first true psychrophilic strains of oscillatorians (isolated from Bratina Island, Antarctica) that have optimal growth at 8°C and cannot survive at temperatures in excess of 20°C. Related Arctic psychrophilic strains were also identified. Phylogenetic analyses of these polar isolates at the ssu rDNA level showed that the few psychrophilic oscillatorians described have arisen in one branch, whereas evolution of the psychrotolerant phenotype has occurred several times (Nadeau et al., 2001). Nadeau et al. (2001) also showed that psychrotolerant strains are most closely related to organisms of temperate latitudes. The occurrence of a shared rare 11-nt insertion in concert with phylogenetic relationships implies that psychrotolerant strains from both Arctic and Antarctic isolates originated from temperate species, whereas psychrophilic strains appear to have arisen independently. A complete genome sequence of a psychrophilic cyanobacterium will allow scientists to establish a database for examining issues of biodiversity, biogeography, and community structure in these important polar mat-forming organisms. Such analyses may also reveal the mechanisms of temperature tolerance of the psychrotolerant species and the mechanisms of low-temperature adaptation of the psychrophilic species. Comparison of the psychrophilic Oscillatoria genome with the genomes of marine cyanobacteria already available may reveal clues as to the factors limiting distribution of the latter group. Understanding the evolutionary relationships of polar mat-forming oscillatorians may have important implications for the study of the origins of life on our planet and others (see Chapter 2). Given the important role that cyanobacteria played in the formation of atmospheric oxygen, knowledge of their phylogeny will also provide new information on the evolution of oxygenic groups and planetary geochemistry.

## Protists

*Polar algae. Chlamydomonas subcaudata* is a green psychrophilic alga isolated from the permanently ice-covered Lake Bonney in the McMurdo Dry Valleys in Antarctica (Lizotte and Priscu, 1992). Because of the ice cover and subsequent lack of vertical mixing, the temperature, nutrient, and irradiance regime experienced by this organism *in situ* is extremely stable. *C. subcaudata* is the dominant species in the deep trophogenic zone (17-20 m) of Lake Bonney, where the average temperature and maximum irradiance during the austral summer are 4 to 6 degrees C and 14 µmol photons $m^{-2} s^{-1}$ (Lizotte and Priscu, 1992). Light penetrating to this depth is mostly in the blue-green wavelengths owing to differential attenuation by the ice cover.

As a psychrophile (Morgan et al., 1998), *C. subcaudata* exhibits unique physiological responses to low temperature when compared to temperate algae or psychrotolerant cyanobacteria isolated from the poles. When exposed to moderate irradiance (150-250 µmol $m^{-2} s^{-1}$) and low temperatures (5-10°C), most species of temperate algae and psychrotolerant cyanobacteria show lower chlorophyll content per unit biomass, smaller amounts of photosystem II harvesting proteins, and increased carotenoids (Maxwell et al., 1994, 1995; Tang et al., 1997b), resulting in a visually yellow or orange color. The adjustments in pigment content and light harvesting capabilities allow the cells to maintain balance between the light energy absorbed through photochemistry and the energy consumed through metabolism (Huner et al., 1998), and they protect the cells from photoinhibition. Unlike the other algae and cyanobacteria, *C. subcaudata* displayed none of these physiological characteristics when grown under moderate irradiance (150 µmol $m^{-2} s^{-1}$) and low temperature (8°C); (T. Pocock, 2002, University of Western Ontario, personal communication).

Compared to a mesophilic species *C. reinhardtii*, *C. subcaudata* had rather low levels of photosystem I (Morgan et al., 1998), indicating adaptation to a predominantly blue-green light spectrum (Neale and Priscu, 1995). Furthermore, *C. subcaudata* possessed high levels of xanthophylls and low levels of β-carotene, suggesting that this phytoplankton species has efficient light harvesting but reduced photoprotective ability compared to *C. reinhardtii* (Neale and Priscu, 1995, 1998). Despite its constant exposure and its photoacclimation to low temperature and low irradiance, *C. subcaudata* retains its capacity to adjust its pigment composition via the xanthophyll cycle, thereby allowing the cell to photoacclimate to high irradiance and to resist photoinhibition (Morgan et al., 1998).

Given its specific adaptation to a narrow spectral distribution and its ability to photoacclimate to low and high irradiance, genomic comparison of the *C. subcaudata* to *C. reinharitii* could further our knowledge of how algal cells photoadapt and photoacclimate to spectral quality and quantity. C. reinhardtii, a temperate algae commonly used as model system is currently being sequenced at Duke University (<http://www.biology.duke.edu/chlamy/index.html>).

*Phaeocystis antarctica* is the primary Antarctic species within the globally important genus *Phaeocystis* and is one of the most important representatives of the family of Prymnesiophytes in the planktonic environment (coccolithophores are the other). *Phaeocystis* occurs in almost every open ocean habitat and forms large blooms in Arctic, Antarctic, and temperate waters (Smith et al., 1991). One distinctive characteristic of *Phaeocystis* is the formation of globular cell colonies, sometimes containing thousands of cells and attaining millimeter size. *P. antarctica* has several characteristics that make it both biochemically interesting and a key organism in Antarctic biogeochemistry. *P. antarctica* forms dense blooms during the Southern Ocean spring-early summer in both pack ice and open water areas (e.g., Ross Sea Polynya; Di Tullio et al., 2000). These blooms provide an important source of early-season organic carbon to these waters, in part because carbon exported from the blooms tends to stay in the water column (DiTullio et al., 2000). *P. antarctica* is also the major planktonic source of organosulfur compounds, which control atmospheric concentrations of dimethyl sulfide (DMS), a climatically important gas (Smith and DiTullio, 1995). One of the current questions in the phytoplankton ecology of the Southern Ocean is what controls the relative abundance of *P. antarctica* versus Antarctic diatoms, which co-occur in spring-summer blooms (Smith et al., 2000). The blooms tend to be spatially separated in the Ross Sea, but the reasons for the spatial separation are unclear. The difference has biogeochemical significance because diatoms and *Phaeocystis* favor different consumer and decomposer assemblages, and diatoms are the source of most carbon buried in Southern Ocean deep-sea sediments (Arrigo et al., 2000; Nelson et al., 1996). Open ocean enrichment experiments have also revealed that diatoms and *P. antarctica* appear to have different responses to limitation by dissolved iron (Boyd et al., 2000).

The response of *P. antarctica* to ultraviolet (UV) exposure is another topic that has received attention from Antarctic investigators. Because *P. antarctica* blooms in the early spring, this species is exposed to solar UV during the period when the development of the ozone hole leads to large increase in UV-B irradiance in the polar biosphere. *P. antarctica*, like other species of the genus, can accumulate high concentrations of the major UV-absorbing compounds, mycosporine-like amino acids (Marchant et al., 1991). These compounds accumulate in larger amounts in this alga (in proportion to other pigments) than in any other Antarctic phytoplankton, making *P. antarctica* a particularly attractive system to study the regulation and function of microsporine-like amino acids (MAAs) in marine phytoplankton (Moisan and Mitchell, 2001; Riegger and Robinson, 1997). Furthermore, the extracellular release of MAAs can be studied in *P. antarctica* in the colonial form. MAAs are found in single cells, but they are also excreted in the colonial matrix material (Marchant et al., 1991).

Sequencing *P. antarctica* will define the evolutionary position of the Prymnesiophytes in general and the evolution of *Phaeocystis* as the only Prymnesiophyte genus common in polar waters, clarify systematics of the genus (Medlin et al., 1994), and enhance our understanding of sulfur cycle and responses to UV radiation. The Arctic congener, *P. arctica*, shares most of the physiological traits of *P. antarctica* but is a distinct species as defined by a number of characteristics, including ssu rRNA sequence (Medlin et al., 1994). Comparing the genomes of those two organisms would provide additional insights into the factors driving phytoplankton speciation.

## Metazoans

*Antarctic fishes: Dissostichus mawsoni, Chaenocephalus aceratus, and D. eleginoides.* Among polar organisms, the phylogenetic history of the Antarctic notothenioid fishes is, without doubt, the most complete (Chen et al., 1998; Eastman, 2000; Eastman and McCune, 2000; Ritchie et al., 1996). Living at constant extreme cold for ectothermic bony fishes required adaptive changes in their biochemical and physiological functions; thus, the notothenioids are a "swimming library" of cold-adapted genes and proteins. We have exciting glimpses of some of these changes: (1) the paradoxical loss of vital cell types, genes, and proteins, including the oxygen-binding protein hemoglobin and red blood cells in the icefish family; and (2) the evolution of novel genes that encode proteins with new functions, exemplified by the antifreeze glycoproteins (AFGPs) of most notothenioids. Currently, laboratories throughout the world are engaged in mechanistic studies of biochemical and physiological adaptation to cold and of the gain and loss of genes, but these efforts are focused largely on discrete traits or gene families.

Sequencing the genomes of three select species of the suborder Notothenioidae (Gon and Heemstra, 1990) could enhance our understanding of environmentally driven evolutionary processes. Two of the three species are endemic to the Antarctic and the other is a cool-temperate relative: (1) the Antarctic toothfish *Dissostichus mawsoni*, a member the oldest lineage (the family Nototheniidae); (2) the Antarctic blackfin icefish *Chaenocephalus aceratus*, a member of the most derived family (the icefishes, Channichthyidae); and (3) the Patagonian toothfish *D. eleginoides* (a cool-temperate congener of *D. mawsoni*). Comparative analyses of these genomes should provide major insight into the progression of evolutionary events that led to the explosive diversification of the notothenioid lineage from its origin as a temperate stock. The haploid genomes of these fishes probably measure ~2 picograms (pg), or approximately two-thirds the size of the human genome. Once one fish genome has progressed sufficiently, the sequencing of subsequent species will be greatly eased by the ability to assemble onto linkage scaffolds established for the first.

*Mammalian hibernators: The Arctic ground squirrel and the black bear.* Several mammals overwinter in extreme conditions by entering a state of suspended animation known as hibernation (Boyer and Barnes, 1999). Although little is known about the molecular genetic events that underlie the hibernating phenotype, the interspersed phylogenetic distribution of hibernating and nonhibernating species has led to the hypothesis that rather than requiring the creation of novel gene products, hibernation results from the differential expression of existing genes. Therefore, it is possible that a small number of genetic events are necessary to acquire the ability to hibernate. The mammalian hibernator genome project would focus on sequencing the genomes of two animals that have different strategies of hibernation. The sequencing work would be complemented by studies of the patterns of tissue-specific gene expression that enable these animals to express the hibernation phenotype.

Arctic ground squirrels (*Spermophilus parryii*) and black bears (*Ursus americanus*) are suitable for elucidation of the genomic and transcriptome-level changes that support hibernation because their hibernation cycles are extremely predictable and physiological changes are so dramatic. They survive the winters of Alaska without eating or drinking for six to eight months by reversibly lowering their metabolism. This metabolic shift has profound ramifications for every mammalian physiological system, yet there are significant differences between the hibernation characteristics of squirrels and bears. Ground squirrels reduce their body temperature as much as 40°C and attain core body temperatures near –2.8°C (Barnes, 1989). Black bears reduce their body temperature by only about 5°C. Ground squirrels lose protein and bone mass, while bears maintain both. A comparative approach to analyzing the genomes and the differences in gene expression patterns in ground squirrels and bears during hibernation will facilitate our understanding of the underlying molecular mechanisms that provide tolerance by molecules, cells, and organs to these extreme changes and will provide great potential for beneficial biomedical applications for humans. For example, understanding the molecular mechanisms of bone mass maintenance in bears may lead to therapeutic modalities that prevent osteoporosis in chronically hospitalized patients (Becker et al., 2002). During hibernation in squirrels, blood flow to the several tissues is reduced by as much as 98 percent of normal for up to three weeks, yet no tissue damage from reduced oxygen availability occurs because the metabolic rate is similarly reduced (Boyer and Barnes, 1999). Identification of the molecular genetic mechanisms affording protection from low blood flow and reperfusion may be applied to protection from injury due to stroke and heart attacks in humans. Another potential use of data obtained in the study of hibernators could be in the development of emergency field medical protocols for inducing a state of hypometabolism in gravely injured humans, for example, soldiers wounded on the battlefield who cannot be transported rapidly to a medical center. By inducing reductions in metabolic rate and enhancing tolerance of reduced blood flow, mechanisms may be developed for sustaining life until sophisticated medical attention can be given to a patient.

*Polar nematodes.* Nematodes in Arctic and Antarctic soils are predators of bacteria, fungi, and other microscopic animals and can be the dominant invertebrates in some polar soil systems. They are important in soil foodwebs because they feed on the primary decomposers (bacteria, yeast, fungi) and influence the rates of decomposition and nutrient cycling.

In the Antarctic Dry Valleys, the bacterial-feeding nematode species *Scottnema lindsayae* lives in the extremely dry soils in water films around soil particles. When unfavorable environmental conditions occur (such as decreasing moisture and temperature), the animals enter into a metabolic state, termed anhydrobiosis (life without water), enabling them to freeze and survive (Treonis et al., 2000). Favorable soil temperature and moisture revive the nematodes, enabling them to save energy for those periods most favorable for activity. The gene for anhydrobiois has been found in temperate fungal feeding nematodes (Browne et al., 2002). Elucidation of the molecular mechanisms for survival of a nematode such as *S. lindsayae*, which occurs in the most extreme soil environment on Earth, will contribute to knowledge of developmental biology and to comparisons with the well-known model nematode, *Caenorhabditis elegans*, which also feeds on bacteria but is not found in polar systems (Freckman and Virginia, 1998; Riddle et al., 1997). *S. lindsaye* thus offers an excellent polar organism for determining and comparing genetic mechanisms of survival to those already elucidated in temperate nematodes.

*Polar insects.* Insects are the most common animals on Earth. Nearly 75 percent of the known species of animals are insects. Furthermore, insects live almost everywhere (except in the oceans), thrive in the Arctic, but are rare in the Antarctic.

The Arctic beetle, Cucujus clavipes, is extremely cold tolerant, with a mean lower lethal temperature of –40°C (J. Duman, unpublished observations). It occurs over a very wide latitudinal range, from Kentucky to Wiseman, Alaska (south side of the Brooks Range). This beetle winters in several larval stages and as an adult. Generally a freeze-avoiding species, *C. clavipes* prevents its tissues from freezing through production of antifreeze proteins. However, the beetles sometimes winter in a freeze-tolerant state, meaning that they can freeze and survive. Clearly, the genetic mechanisms for avoidance and tolerance of freezing may be elucidated by genome analysis, most likely at the level of the transcriptome. Furthermore, Alaskan populations winter in a deep diapause state, whereas those in Indiana do not. *C. clavipes*, therefore, provides a model system for genetic dissection of diapause as well as survival of metazoan tissue during freezing.

### Plants

*Betula nana*. The dwarf or bog birch is one of the most characteristic plants of the low Arctic region. It is found around the world and is the dominant plant in many areas. In other areas, such as Alaskan tussock tundra, it remains an important secondary species. Since the extent of shrub cover is critical in controlling snow distribution, *Betula nana* affects many aspects of the biophysics and the climate dynamics of the Arctic. In North America, it is very responsive to environmental manipulation and changes such as increased nutrients or warming. Experimental warming experiments (in small, *in situ* greenhouses) can produce a small forest of *Betula nana*. However, in Scandinavia, the same species appears much less responsive to nutrient additions. Given the importance of *Betula nana* in Arctic ecology and climatology, it is important to understand its physiology and its range of responses. Furthermore, it may be important to understand the nature of genetic variation that exists around the Arctic world.

### Other Considerations

The polar species cited above, based on their biology, represent examples of compelling opportunities for genome science projects. We emphasize that not all projects will require the sequencing of the complete organismal genomes. Depending on the scientific questions and objectives, many projects may be addressed more effectively, and with greater cost efficiency, by other genome-wide methods (transcriptional profiling, protein gel profiling). Hence, it will be necessary to develop a framework for prioritization of polar organisms for full genome sequence characterization versus functional genomic profiling. Given the small sizes of most prokaryotic and many protistan genomes, they can be sequenced to completion provided that a strong scientific justification is advanced. The large genomes of metazoans and plants, by contrast, will require careful assessment of the scientific benefit versus programmatic cost. One possible scenario for initiating a Polar Genome Science Initiative is outlined here:

1. Full-scale genome projects can be launched for one or two metazoan or plant species whose biology is well understood.
2. Meanwhile, 8-10 other animals and plants would be selected for functional genomic analysis via the construction of EST (expressed sequence tag) libraries, microarray production, and proteomic profiling. The results of the functional genomic studies should indicate whether these genomes deserve more detailed study, and 8-10 EST/proteomic projects could be executed for the price of one complete genome project. Gene expression or protein turnover profiles exhibiting potentially "adaptive" features would argue for advanced analysis of the appropriate genomic regions, which could be cloned out of BAC, PAC, or YAC libraries (see below). Furthermore, the development of specific hypotheses based on the functional genomic approach would naturally define the appropriate comparative taxa while generating economy in focusing the work.

The importance of "testing" putative environmental adaptations within the genes and genomes of polar organisms by comparison to phylogenetically related, but temperate, species (criterion 3) cannot be overemphasized. A Polar Genome Science Initiative, by its very nature, will require a strong comparative genomic component; and suggestions for appropriate species comparison are given in previous sections. Finally, as the initially exploratory phase of these genomic projects proceed, we anticipate that each will transition to directed, hypothesis-driven research based on the discoveries made in the first phase. Rigorous analysis of adaptation using approaches such as phylogenetically independent contrasts (Felsenstein, 1989) will be necessary to distinguish adaptive variation from the influences of ancestry.

Because the generation times of most polar organisms are so long, none are likely to be developed into genetic "model organisms." Thus, the functional attributes and/or biotechnological potential of a gene obtained from a polar species must be assessed by reverse genetic strategies (e.g., manipulated expression of the gene by antisense morpholino RNA oligonucleotide "knockdown" [Nasevicius and Ekker, 2000] or by gene transfer methods, etc.), perhaps conducted in the organism itself or, more likely, in conventional model systems amenable to such approaches (e.g., various bacterial species, the plant *Arabidopsis thaliana*, the nematode *Caenorhabditis elegans*, the zebrafish *Danio rerio*, or the mouse *Mus musculus*).

## STRUCTURE OF A POLAR GENOME PROJECT

Whatever the organism under consideration, a genome project will include the following aims (Gibson and Muse, 2002), some of which may take precedence depending on the scientific objectives, the biological characteristics of the organism, and available resources.

### Generation of Physical Maps of the Genome

The genes within a genome are arrayed along one or more chromosomal DNA molecules, which may be circular (common in prokaryotes) or linear (typical of eukaryotes). The location of the genes may be specified by relative position determined by recombination frequency (a genetic map) or by physical distance along the DNA (a physical map). Given the long generation times of many polar organisms (especially metazoans), mapping genes with respect to genetic markers through controlled crosses is unlikely to be feasible. Similarly, the absence of sexual conjugation systems for polar prokaryotes precludes genetic mapping strategies in these organisms. Fortunately, the development of several physical mapping techniques that assemble contiguous segments of DNA (*contigs*) based on landmark sequences (such as sequence-tagged sites, [STSs]) provides practical alternatives to genetic mapping. The ultimate physical map for an organism is the complete sequence of its genome, although one must recognize that individual genomes may vary due to indel, inversion, and single-nucleotide polymorphisms (SNPs).

The physical mapping of the genome of a polar organism will undoubtedly require a combination of techniques, including the following:

- restriction fragment length profiling, in which large clones of the organism's DNA (100-1,000 kilobases [kb]) randomly selected from YAC, BAC, or PAC libraries are aligned based on shared restriction fragments (Gibson and Muse, 2002);
- chromosome walking, in which the sequences of the ends of one fragment are used as hybridization probes to identify overlapping clones (Gibson and Muse, 2002);
- radiation hybrid panels, wherein hamster fibroblast cell cultures that harbor fragments of the organism's chromosomes are probed by species-specific polymerase chain reaction (PCR) and the frequency of cosegregation of two genes provides an estimate of the distance separating them (Deloukas et al., 1998; Kwok et al., 1999; Walter et al., 1994); and
- whole-genome shotgun assembly, in which genomes are fragmented into small pieces (2-5 kb), the ends of the clones are sequenced,

and the fragments are assembled computationally into scaffolds (cf. the *Fugu* genome; Aparicio et al., 2002).

## High-Throughput Sequencing of Genomic DNA and Expressed Genes

The *sine qua non* of a genome project is the ability to sequence rapidly the fragments of a genome, whether large or small, with sufficient redundancy (six- to tenfold coverage) to reduce the error rate to between 1 in 1,000 nucleotides (a "rough-cut" genome) and 1 in 10,000 (a "polished" genome). Today, the most common sequencing method is based on automation of the Sanger dideoxynucleotide chain termination protocol (Sanger et al., 1977). If large-insert clones have been used to establish the physical map, then the clones must be subdivided to produce pieces (~1-2 kb) amenable to sequencing. Thus, one advantage of the shotgun mapping strategy is that libraries of short fragments are the starting point. Once sufficient numbers of sequenced fragments have been obtained, they are ordered into contigs and the contigs into larger "scaffolds" of genome sequence. The sequences of expressed genes (for example, cDNAs, or complementary copies of messenger RNAs [mRNA]) are also incorporated into the assembly because they help to define the intron and exon boundaries of the genes in the genome. Irrespective of effort, some genomic regions will be refractory or "unsequenceable." Often these regions have a biased, high guanine-cytosine (GC) content or consist of short repetitive elements that are difficult to resolve. Whereas microbial genomes are generally sequenced to completion, the "finished" genomes of eukaryotes will normally contain gaps.

A major consideration for any genome project, such as the Polar Genome Science Initiative contemplated here, is the cost of the sequencing itself as well as the computational power required to assemble the genome. Fortunately, new sequencing technologies promise to reduce costs to levels unimaginable at the start of the public and private sector human genome projects. The National Human Genome Research Institute has just funded GenomeVision to reduce the costs of large-scale gene sequencing projects by five- to tenfold through miniaturization over the next two years (GenomeWeb staff, 2002). Many alternative technologies are being developed that should increase the speed and accuracy of sequencing while lowering costs (Lakhman, 2002; McGowan, 2002a). Thus, the Polar Genome Science Initiative is not only intellectually compelling but also imminently practical and affordable.

## Gene Identification and Annotation

The genome of the pufferfish, *Fugu rubripes*, is estimated to contain ~31,000 genes, or roughly the same number as current estimates of the human genome (Aparicio et al., 2002). Of predicted human proteins, ~75 percent are orthologous to pufferfish proteins, whereas the remaining 25 percent either are highly divergent or are not encoded by the fish genome. This comparison emphasizes that gene prediction must be pursued both by orthology and by use of *ab initio* gene prediction tools. Following identification, genes must be annotated with data regarding presumptive function, pattern of expression, and putative orthologues found in other genomes.

## Population Analysis with Single-Nucleotide Polymorphisms

Generating a genome sequence based on one, or at most a few, individuals of a species represents merely a beginning for population biologists. Natural variation among genomes in a population is the "stuff" of phenotypic variation and evolutionary speciation. It is generally assumed that single-nucleotide polymorphisms and indels are responsible for quantitative variation in phenotypic traits. SNPs may be used to track gene flow between separate populations of a species, and their absence signals that the populations are stratified and perhaps in the process of speciating due to ecological, geographic, or behavioral factors. Because DNA-sequencing costs are declining rapidly, the identification of robust "SNP libraries" for population analysis of multiple species is a realistic goal.

The capacity to compare distinct populations of a polar species and to monitor community relationships between interacting species using SNP technology promises to revolutionize polar ecology. Some potential applications (Gibson and Muse, 2002) include:

- inference of the demographic history of populations;
- analysis of mating systems;
- conservation biology, including the population forensics of commercially exploited species;
- analysis of breeding structure and dispersal of soil microorganisms, nematodes, and so forth; and
- timing the establishment of host-symbiont and host/parasite associations.

## Web-Based Databases and Interfaces for Data Management and Comparative Genomics

Genome projects produce massive amounts of sequence data and annotated information. These data must be made available to the wider biological research community by creation of appropriate relational databases and web-based interfaces. Indeed, the ability to compare genomes will speed our understanding of genome evolution and the phylogenetic relationships of all organisms, whether polar or temperate. A comprehensive Polar Genome Science Initiative must make provision for creation, curation, validation, and management of these databases and for the bioinformatics tools necessary for insightful genome analyses.

## Transcriptome Analysis

As emphasized at several junctures in this report, the ability to qualitatively and quantitatively describe the transcriptome opens up a number of new avenues for investigating polar organisms. All taxa can be examined through transcriptome analysis, and studies can involve complex microbial consortia as well as individual animals or plants and tissues thereof. A primary use of transcriptome analysis is to study how environmental factors, both singly and in combination, influence patterns of gene expression. The environmental factors of interest comprise natural variables such as temperature and UV radiation and anthropogenic factors such as organic and heavy metal pollutants. Transcriptome analysis can provide a "snapshot" of the organism's status in terms of gene expression and makes it possible to follow the time course of organismal responses to environmental change.

Although the use of DNA microarrays for examining organisms' transcriptional responses to the environment is in its infancy, there are several indications of how promising this approach can be for probing the effects of environmental change. Studies of yeast have shown that a characteristic set of stress-related genes is activated upon exposure of the cells to a variety of stresses (anoxia, temperature, alcohols, and so on) (Causton et al., 2001; Gasch et al., 2000). Stress-specific alterations in gene expression were also catalogued in yeast. This technology is becoming accessible to scientists interested in all types of organisms, from model systems to species for which no sequencing of the genome has been done (Pennisi, 2002).

The fabrication of DNA microarrays for transcriptome analysis can involve a number of experimental strategies. For organisms having a fully sequenced and well-annotated genome, DNA "microarrays" fabricated with specific oligonucleotide probes for the gene (mRNAs) of interest can be built. Customized "microarrays" are available from a number of commercial sources, and this type of commercially produced technology will certainly become increasingly available for transcriptome analysis of many different organisms. In the case of species for which sequence information is limited or even entirely lacking, the construction of DNA microarrays must follow a different strategy. To construct microarrays for non-sequenced species, strategies such as that described by Gracey et al. (2001) are likely to be effective. Through construction of subtracted and normalized cDNA libraries, thousands of different cDNAs for spotting onto microarrays can be obtained. Through iterative analysis of these microarrays, one can screen the cDNA libraries to obtain thousands of unique cDNAs with minimal redundancy. Techniques are also well developed for selecting for full-length cDNAs so as to increase the utility of the cDNAs produced in microarray studies. Although DNA microarrays fabricated for "nonmodel" organisms offer an effective means for screening changes in gene expression, they have two key limitations. One stems from the fact that these arrays contain an incomplete representation of the genome. The second is that the absence of extensive sequence information limits the identification of many expressed genes. Also, the usefulness an array constructed for one species in study of another remains to be determined. This is an important question for future work. Despite their limitations, DNA microarrays for "nonmodel" species offer a powerful tool for analyzing the effects of environmental factors on gene expression.

### Proteome Analysis

Changes in the transcriptome do not map one to one with changes in the proteome (Fiehn, 2001; Phelps et al., 2002). Thus, depending on the goals of a study, analysis of the transcriptome may serve as only an initial step in the study of how environmental changes affect the phenotype. Proteomics is a powerful approach that allows one to characterize the suite of proteins present in a cell or tissue.

The applicability of proteomic methodologies to the study of polar organisms, for which large amounts of DNA and protein sequence data are not available, appears promising for several reasons. First and foremost, the conservation found in the sequences of orthologous proteins facilitates identification of proteins from genetically uncharacterized species. Second, as more and more genomes are sequenced and increasing amounts of information are obtained on the deduced amino acid sequences of orthologous proteins, proteomic analysis of nonsequenced organisms will become increasingly feasible. Targeted proteomics, in which only a minor fraction of the proteome is analyzed, for example, using antibody methods, may be the most suitable strategy for screening changes in the levels of proteins that are of interest in a particular physiological context. Analysis of the transcriptome may point to the set of proteins that are most important for proteomic analysis.

## Metabolome Analysis

Characterization of the composition of metabolites in the cell—*metabolomics*—carries analysis one step closer to actual physiological activity (Phelps et al., 2002). Metabolome analysis allows the charting of the types of substrates, end products, and biosynthetic intermediates found in the cell. Through appropriate coupling of analytical technologies, metabolomic approaches can be quantitative as well as qualitative. With the advent of protocols in which magnetic resonance spectroscopy is coupled with effective separation techniques and mass spectrometry, identification and quantification of virtually all organic molecules in the cell is becoming possible (Fiehn, 2001).

Like the analyses of the transcriptome and the proteome, characterization of the metabolome offers enormous potential for discerning the effects of environmental factors on organismal function. Similar to transcriptome and proteome analyses, metabolomic methods can be applied to any type of organism and to different cell types and tissues within an individual.

## Ecogenome Analysis

Ecogenomics, the use of genome science to study ecology, has great potential to advance our understanding of microbial ecology (Stahl and Tiedje, 2002; Torsvik and Ovreas, 2002). One promising approach is metagenome analysis (Rondon et al., 2000). This approach is based on the same technology that is used to sequence the whole genome of specific organisms, but it is applied to entire microbial communities. In these analyses, large DNA fragments are extracted directly from microbial communities, large extents of sequence are determined, and the sequences are partially analyzed. In theory, the whole genomes of members of the community sampled can be assembled. From metagenome analysis, the following information (at a minimum) can be gleaned: phylogenetic composition of the sample, variability of recognizable functional genes, association of functional genes with a phylotype, indications of new and unsuspected functions, dosage of a particular gene in a chromosome or contig of interest, and insights into the regulation of gene expression. One important task for investigators of ecogenomics is to develop means for studying both culturable and unculturable (at least by present technologies) species, the latter often representing >99 percent of a microbial community. Thus, microarrays developed for examining community structure and function must include probes from both culturable and unculturable species.

Although metagenome analysis is still in the development and testing stage, it holds great promise for providing a new, integrated view of the phylogenetic composition of microbial communities and of their functional capabilities. To date, only a few studies have employed metagenomic analyses (Béjà et al., 2000; Rondon et al., 2000). Ambitious plans for more such studies have been announced (McGowan, 2002b). Perhaps the greatest potential of such studies lies in their ability to address the critical need to relate microbial phylogeny to function. To some extent, this has been and can further be accomplished by determining linkages between "functional genes" and indicators of phylogeny such as rRNA genes.

The metagenome approach may be particularly appropriate for polar problems. First, contig assembly is simplified if simple rather than complex communities are studied. Simple communities may be expected in some of the more extreme polar environments, for example, wintertime sea-ice communities, Dry Valley soils (or their Arctic equivalent), lake ice bubbles, and possibly subglacial lakes (see Chapter 2). Second, metagenome approach could yield information about the composition and functioning of microbial communities that are particularly difficult to sample without disturbance or that are not amenable to experimental manipulation, such as subglacial lakes and sea-ice microbial communities.

By analogy to single-organism genomics, ecogenomics must make the transition from sequencing and annotating metagenomic data to functional analyses. By further analogy to single-organism genomics, several genomic approaches appear to hold promise for functional ecogenomics. "Environmental microarrays" have several potential applications. Measurement of the dynamics of large numbers of individual populations may be possible using probes for indicators of phylogeny. Population dynamics can then be related to environmental data. Similarly, estimation of the abundance of large numbers of genes with known function in communities allows population studies of microbial guilds (populations sharing a common function in a community). Furthermore, using environmental microarrays, it may be possible to do transcriptional analysis, permitting estimates of *in situ* activity of functional guilds. Complementary to transcriptional analysis, "environmental proteomics" may provide an additional approach to estimating the *in situ* activity of guilds. Moreover, "environmental metabolomics" may provide a third approach for estimating *in situ* activities of guilds, which would not be limited by our genetic knowledge of the organisms in a community. These functional genomic approaches are applications based largely upon metagenomic analyses and should be closely coordinated with the latter analyses. Relevant metagenomic data should be readily available to facilitate the application of the functional genomic approaches. Finally, bioinformatic approaches may determine relationships between phylogenetic groups and measurable functions, particularly if a common database is established relating ecogenomic data to phenotypic, geographic, and environmental data (Stahl and Tiedje, 2002).

All of these functional ecogenomic approaches involve severe technical challenges, and none has yet been satisfactorily demonstrated. Several groups are actively developing various types of environmental microarrays (Guschin et al., 1997; Small et al., 2001; Wu et al., 2001). Progress has also been made in environmental transcriptional analysis (Bakermans and Madsen, 2002; Miller et al., 1999; Park et al., 2002). Environmental proteomics and metabolomics are currently hypothetical approaches. All of these approaches must address the extreme complexity of DNA, RNA, and proteins in most environments, which tends to increase detection limits and decrease specificity of analyses. Another problem common to environmental samples is the complexity of the sample matrix, which can limit analysis recovery and interfere with analyses. Despite the challenges, the great potential of these ecogenomic approaches merits exploration. Leading microbial ecologists have endorsed ecogenomic research (Stahl and Tiedje, 2002), and application of ecogenomics to marine microbial ecology has been recommended in a previous report (NSF, 2000). Ecogenomics has great potential for addressing some of the research questions in polar biology outlined in the previous chapter.

### Impediments to the Study of the Transcriptomes, Proteomes, and Metabolomes of Polar Species

Implementation of the study of transcriptomes, proteomes, and metabolomes of polar organisms faces a number of challenges, most of which are common to all three types of "-omic" analysis. In each case, the equipment needed to conduct this research is expensive and requires skilled hands for its operation. In the case of transcriptome studies, the equipment needed to fabricate and analyze DNA microarrays—for instance, robotic apparatus for spotting DNA onto slides and for handling large numbers of liquid samples—is very costly and it is not likely that all research centers will be able to acquire this equipment. Therefore, efforts should be made to provide access to the technology needed for transcriptome analysis for scientists working at sites where technology shortfalls exist. Identical arguments apply in the case of the equipment required for proteomic and metabolomic studies. When the required equipment is present at a center, it is likely to be housed in a central facility for use by multiple investigators. Technical support for equipment operation and maintenance will likely be required at these centers. These technological demands for metabolomic and proteomic studies apply not only to polar science but to other bioscience disciplines as well. Support for a central metabolomic and proteomic center by funding agencies will benefit a broad community of investigators, including polar biologists.

Large amounts of DNA sequencing accompany transcriptome analysis, and facilities for this purpose must be available to investigators. For DNA microarrays spotted with uncharacterized cDNAs, the spots that exhibit interesting patterns of up- or down-regulation must be sequenced to identify the genes undergoing shifts in expression. Sequencing may also precede the construction of microarrays to enable genes of interest to be included on the arrays. Sufficient support for DNA sequencing at research universities may already exist in most cases, so hurdles to implementation posed by sequencing capacity may be relatively small. Furthermore, where sequencing potential is not found, investigators may be able to farm out the needed sequencing to commercial firms or to universities that perform sequencing on a recharge basis.

A final aspect of "-omic" research that merits emphasis is the likelihood that most aspects of these studies will be difficult, if not impossible, to carry out at remote field sites. It seems impractical, for example, to site sophisticated robotic systems for preparing DNA microarrays or equipment for mass spectrometric or magnetic resonance experiments in proteomics and metabolomics at field sites. Instead, what should be guaranteed to investigators is the technology needed for sample preservation (for example, liquid nitrogen or dry ice) and reliable transportation of samples from the field back to the home laboratory where the sophisticated "-omics" analysis will be conducted.

## Bioinformatic Tools and Databases

Another common requirement of "-omics" research is expertise in bioinformatics. The software needed to organize and to analyze the huge sets of data generated in all types of "-omic" studies is often available on web sites at no cost to the user. However, given the interest in looking for mechanisms of environmental adaptation in a given polar species' genome sequence, current tools are not likely to be appropriate for the task. An initiative in polar environmental genomics requires the design and development of specific bioinformatics tools that would search for sequence data that would support or refute hypotheses regarding adaptive processes. Therefore, collaborations between polar biologists and computational biologists and bioinformaticists are essential and should be encouraged.

Fellowships may be set up to fund students and postdoctoral researchers in the computer and mathematics departments to participate in polar genome studies.

As genomics-based efforts in polar biology research expand, sufficient attention must be devoted to the bioinformatics issues related to the distribution and use of the information. Genome sequence data alone are relatively easy to access; however, the diversity of polar organisms being considered for whole-genome analysis (microbial species and various eukaryotes and metazoa), together with the desire to link genome sequence data to geographical and environmental (geochemistry and climatic) data and temporal data, presents a much greater bioinformatics challenge. This type of data sharing and linking requires the development of a fully integrated database for which no robust model yet exists. Moreover, investigators must be willing to agree on a set of standards for data format and data sharing. Such large matrix arrays of data can only be fully analyzed using network structure models. Maintenance of such databases and development of network analysis tools would require long-term funding, quite likely from multiple funding agencies. The need to develop integrated databases to link genome sequence, function, ecological, climatic and geographical, and temporal data is not unique to the polar research community. The polar research community should become actively involved in the ongoing database discussions that are taking place.

## CREATION OF A POLAR GENOME SCIENCE INITIATIVE

Given the great potential of genomic science to address important new research questions in polar regions, some special effort to facilitate and guide these activities is justified. The goal of such an initiative would be to gather talent to work on these problems in an efficient and coordinated manner.

One option for a Polar Genome Science Initiative would be for the community to form some kind of virtual steering committee or core group to provide leadership, perhaps based on the model offered by the International Arctic Polyna Programme (IAPP) of the Arctic Ocean Sciences Board (AOSB). The AOSB is a nongovernmental body that includes members and participants from research and governmental institutions. Its long-term mission is to facilitate Arctic Ocean research by supporting multinational and multidisciplinary natural science and engineering programs. In doing so, it encourages communication, promotes information exchange, and facilitates discussions of needs and priorities. The IAPP Science Coordinating Group comprises volunteer scientists who serve to define the scientific needs and to coordinate the execution of research, and the AOSB serves primarily to facilitate discussion and build networking opportunities. Part of the success of AOSB is the ability to build international cooperation on what is an inherently international topic, the Arctic Ocean. AOSB members come from Canada, Denmark, Finland, France, Germany, Iceland, Japan, the Netherlands, Norway, Poland, Russia, Sweden, Switzerland, the United Kingdom, and the United States of America. Although this international emphasis is not necessary for the Polar Genome Science Initiative, the model of an informal collaborative body might be useful.

The main advantage of this approach is that it is relatively inexpensive (although there are still costs associated with supporting a secretariat and quality web presence, and there are costs incurred directly by each participant for travel). The main disadvantage is that this approach may not necessarily be able to facilitate implementation of the steering committee or core group's thinking without the ability to leverage its ideas and plans (with funds) into activities. It requires a significant amount of effort from its volunteer participants, so a core group of truly interested leaders must emerge and be active if it is to make progress. If the Polar Genome Science Initiative is modeled after AOSB's IAPP within the United States, concerns might be raised as to why this informal group has the credibility to "speak" for the discipline. (For more information on the AOSB, see <http://www.aosb.org/>.)

In the committee's opinion, a more effective approach for a Polar Genome Science Initiative would be for the National Science Foundation (NSF) to consider this recommendation as a priority area, providing targeted funding and facilitating establishment of a Science Steering Committee to lead the planning. This approach might be modeled after the Arctic System Science Program's (ARCSS) Ocean-Atmosphere-Ice Interaction program (or similar ARCSS's programs). The Steering Committee of the Polar Genome Science Initiative would include representatives of the relevant biological communities and would meet periodically to do strategic planning, set research priorities, discuss needs and how they might be met, solicit further input from the broad biological community, and encourage coordination and communication.

Using this approach to implement a comprehensive, coordinated Polar Genome Science Initiative would generate synergies of effort that would maximize scientific output while minimizing the resources required. Under this approach, the Scientific Steering Committee would establish priorities and coordinate large-scale efforts for genome-enabled polar science (for example, genome sequencing, transcriptome analysis, coordinated bioinformatics databases). There would be no immediate need for new facilities or capabilities; instead, the initiative would support "virtual polar" genome science centers, recruited by NSF from the many extant genome centers, to provide the equipment and expertise necessary to support the polar biological community. Other advantages of this strategy include:

- pooling of resources for analysis of multiple genomes, with concomitant economies of scale;
- division of labor to enhance the efficiency of the scientific return;
- coordination of community efforts to avoid unnecessary duplication of research; and
- provision of uniform databases that facilitate cross-organismal and interpolar genomic comparisons.

The committee believes that this approach is the most effective way to move forward, and the *Arabidopsis* Genome Initiative (see Chapter 5) shows that a well-planned effort can actually finish ahead of schedule if tasks are delegated effectively. This approach also makes it easy for new scientists to participate, because there is a clearly articulated way to engage the process, make contacts within the network, locate information, and seek potential research partners from other fields. This approach could be designed to encourage partnerships between universities and the private sector. The main disadvantage of this approach is that it requires new funding; the polar science community would not likely support it if it meant taking funds away from existing initiatives.

The committee believes that NSF is well positioned to be the lead in the Polar Genome Science Initiative. NSF is the nation's preeminent organization for the support of basic science and the one government agency with the scope and expertise to foster this type of effort in polar science. NSF is a major supporter of research in the Arctic and the key provider of support (with minor other inputs from NASA and others) for activities in the Antarctic, so it is already the acknowledged leader in advancing polar science. In addition, NSF has been funding genomic and integrative biological research through its current programs on Frontiers in Integrative Biological Research (FIBR) and Genome-Enabled Environmental Sciences and Engineering (GEN-EN). However, the suggested Polar Genome Science Initiative is a large-scale research effort that aims to facilitate the application of genome research in the polar regions, coordinate the sequencing efforts of polar organisms, and encourage collaboration between polar and nonpolar scientists. It is beyond the scope of FIBR and GEN-EN. Together, its Office of Polar Programs and Directorate for Biological Sciences already have the expertise necessary to start and manage this kind of initiative, and they have contact with relevant communities to facilitate the transfer of knowledge that would be a key component of the initiative.

# 4

# Complements to Genome Science: Enabling Technologies, Facilities, and Infrastructure

Genomic methods (broadly defined) stand to make significant contributions to polar biology, but the successful application of these new methodologies is likely to depend on a suite of other enabling technologies. Enabling technologies include sophisticated methodologies for integrating biological data with information from geochemical and geophysical analyses, improved means for providing access to field sites during winter conditions, improvements of laboratory and storage facilities, advanced technologies to allow remote sensing of organismal activities, and better methodologies for accessing specimens found in difficult-to-sample habitats such as subglacial lakes.

## ENABLING TECHNOLOGIES

### Examples of New Approaches to Link Organismal and Process Data

New and improved stable isotopic techniques, used alone or in parallel with molecular methods, can help unravel how organisms respond to the uncertainties of global change and how they contribute to the functioning of polar ecosystems. Advances in stable isotope techniques can help scientists understand the links between organisms and geochemistry (e.g., nutrient, carbon, silica, hydrology cycles). Techniques include the following:

- *Multiple element ($^{30}Si/^{28}Si$, Ge/Si, $^{15}N$, $^{13}C$, $^{18}O$) or coupled analyses ($^{13}C$, $H_2^{18}O$).* These analyses improve our ability to study organisms as participants in complex interactions. For example, these analyses can be used to resolve whether plant stomatal or photosynthetic processes are affected by environmental perturbations, or to characterize the $\delta^{18}O$ values of precipitation (soil water or snow) or soil and organism $\delta^{13}CO_2$ under a suite of climate change scenarios, indicating changes in photosynthetic pathways, sources of respired substrates, or other geochemical processes.

- *Compound-specific stable isotope studies.* Compound-specific stable isotope measurements improve resolution of bulk carbon characteristics because the multiple sources of carbon (undegraded plant lignin, polysaccharides, etc.) can be identified (Neff et al., 2002). When compound-specific stable isotope techniques are combined with traditional bulk stable isotope or radioisotope analyses, the power of biogeochemical research increases substantially due to our enhanced ability to resolve carbon sources, including paleocarbon sources.
- *Increased resolution of mass spectrometers.* The high resolution of mass spectrometers now allows minute quantities (<20 µg) of microscopic invertebrate species to be analyzed. The finer resolution increases the potential for integrating the details of food sources, energy, and nutrient transfer in smaller organisms of the microbial food web that were previously neglected or grouped with larger organisms or sediments.

Stable isotope techniques and molecular techniques can be used in parallel to discriminate the ecosystem function of a particular group of microorganisms. Techniques include the following:

- *Functional-level analysis.* Functional-level analysis combines $^{13}C$ tracer studies with specific phospholipid fatty acids (PLFAs). Distinct PLFA profiles are characteristic of such organisms as fungi, Gram-positive bacteria, and actinomycetes. When measurements of PLFAs are combined with stable isotopes, a powerful tool is available for following carbon flow through microbial communities.
- *Stable isotope-probing (SIP) techniques.* SIP is an advanced culture-independent technique that allows isolation of DNA from microorganisms at a species level. An advantage of the technique is that it allows isolation of entire genomes of, for example, active methylotrophs in $^{13}C$-DNA fractions, which enables a parallel analysis of functional gene sequences within populations. These rapidly developing techniques have enormous potential for understanding diversity and function across spatial and temporal scales, as well as for understanding relationships to complex biophysical properties found in polar regions.

## Fluorescence-Based Technology

Fluorescence-based technology, such as fast-repetition-rate fluorometry (FRF) (Behrehfeld et al., 1996; Kolber et al., 1994), pump during probe (PDP) (Falkowski and Kolber, 1993), and pulse amplitude-modulated (PAM) fluorescence (Renger and Schreiber, 1986; Schreiber et al., 1994), is now commonly used to study photochemical yield of algal photosynthesis. The following are some useful advancements in fluorescence technology that may be of use to polar scientists:

- *Microfluorometer.* Measurement of the quantum yield of photosynthesis for single cells or colonies is now possible by attaching an epifluorescence microscope to the PAM fluorometer. Such measurements allow assessment of a single cell or single species' response to different light, nutrient, or temperature regimes and may lead to predictions of taxonomic succession under changing environmental conditions.
- *Submersible PAM fluorometer.* A submersible prototype of the PAM fluorometer that has substantially higher sensitivity than the commercially available dive PAM is being tested in the Ross Sea (W.O. Smith and J.A. Peloquin, personal communication, October 16, 2002). The submersible prototype can be lowered to approximately 30 m, thereby allowing measurements of the vertical structure of fluorescence characteristics. These instruments can be used to study the effects of such abiotic factors as ultraviolet radiation and temperature on the vertical structure of phytoplankton.

## Biosensors, On-board Instrument Packages, Electronic Tagging, and Nanotechnology

Obtaining data on free-ranging organisms in the field is challenging in all environments. However, the polar regions pose special challenges, notably those associated with efforts to observe organisms in the field on a year-round basis. Many of the problems created by the harsh environment can be reduced through the development and exploitation of technologies that involve remote sensing, on-board instrument packages, electronic tagging, and newly emerging applications of biosensors and nanotechnology.

*Instrument packages and tags.* Small and versatile instrument packages can be attached, permanently or reversibly, to study animals that are subsequently released back into the field. These instruments allow a variety of measurements to be made, including geoposition, depth, heart rate, and, in some applications, blood chemistry. (See <http://www.toppcensus.org/proj_plan_workshop_1.html> for a review of this technology and a comprehensive bibliography of studies in which this technology has been employed.)

*Nanotechnology and biosensors.* New types of instruments employing sophisticated DNA and protein-based biosensors that exploit the potential of nanotechnology are in development (see Chibber et al., 2002, for an example). Of special promise are instruments able to sense specific types of biochemical signals in the environment. For example, sensors to which specific DNA molecules or antibodies are attached could provide means for characterizing the compositions of aquatic microbial ecosystems or for charting plankton blooms.

## Subglacial Lake Exploration

The discovery of subglacial lakes some 4 km beneath the surface of Antarctic ice sheets offers unparalleled opportunities for research, but also unparalleled challenges. A number of enabling technologies will be required to study the form, distribution, and activity of life in the lakes (SCAR, 2001). One critical aspect of subglacial lake exploration is testing, verification, and monitoring for potential contamination during all phases of the scientific program. All of the methodologies employed, from ice drilling to sample recovery, must be scrutinized carefully and deliberately from both an environmental stewardship and a scientific standpoint.

To develop a subglacial lake exploration program and move ahead with ice-based biological research, enhanced field logistics and new tools for sampling and archiving data would be needed. Examples include:

- *Fast-access drilling technology.* Subglacial lake exploration will require targeted sampling and observations from multiple boreholes drilled over a wide geographic region. Mobile and rapid drilling is required for such work. One example of fast-access drilling was a hot-water system used until recently to investigate the controls on fast-ice flow in West Antarctica. Unfortunately, this hot-water drill can drill only to a depth of ~1 km, whereas most of the Antarctic ice sheet is considerably thicker (~2-4 km). Clow and Koci (2000) have proposed a system based on coiled tubing technology that illustrates one approach for fast and mobile drilling to all ice sheet depths.
- *Clean drilling and sampling technologies.* These technologies form perhaps the largest hurdle in subglacial lake exploration and must consider both forward and backward contamination issues. Before sampling can begin, a water lock must be developed that will allow all sampling equipment to be properly decontaminated and to ensure that the lakes remain isolated from the atmosphere. The waterlock system should be

integrated with the developing fast access drill technology. Once developed, such technology will open the door for the placement of *in situ* observatories and sample return missions.

- *In situ observatories.* An initial step in subglacial lake exploration should include the deployment of an observatory or series of observatories. These observatories would acquire a time series of such basic measurements as dissolved oxygen, conductivity, redox, pressure, temperature, and turbidity using sensor strings located at selected depths throughout the water column. Molecular arrays should be developed to give an initial picture of the microbial types and their metabolic capabilities.
- *High-pressure culture technology.* Although it is not known if microbes living in subglacial lakes are adapted to elevated hydrostatic pressures and require these conditions for optimal growth, attempts should be made to determine whether experiments with the subglacial lake biota will be enhanced by high-pressure culture technology, as has been used in studies of deep-sea microbes (Yayanos, 1986).

## FACILITIES AND INFRASTRUCTURE

### Creating a Virtual Sequencing Facility to Coordinate Large-Scale Sequencing Efforts

Realizing the full benefits of genomic-based approaches to the study of polar biology will require significant work and sophisticated facilities. A number of facilities capable of supporting sequencing activities for polar biology already exist, along with a large pool of talented scientists, so there is not an immediate need for new facilities. Instead, one of the greatest challenges in the implementation of a genomic research program in polar biology will be to identify the most appropriate mechanisms for accomplishing the work.

Different approaches are possible. At one extreme would be a model where individual investigators with interest in a particular organism carry out genome sequencing activities in their own laboratories. At the other extreme would be the model in which large-scale sequencing centers that routinely produce billions of base pairs (bp) of DNA sequence each year would perform the work. There are pros and cons to both models; however, there is an economy of scale in both time and money that comes from having this work carried out in large centers. In the United States alone, the estimated combined DNA sequencing capacity in medium to large centers is approximately 500 million lanes per year—more than enough to meet the needs of the polar biology research community in the near future. In addition, the large sequencing centers also have the appropriate bioinformatics tools and pipelines to support genome annotation and analysis. What the large centers do not have is the expertise to link DNA sequence data to the biology of polar organisms.

This means that there is no immediate need to build additional infrastructure to support sequencing activities. Instead, the expertise of scientists from diverse backgrounds must somehow be coordinated in some sort of "virtual" sequencing facility. This approach would blend the expertise already within the polar research community with the expertise in genomics and bioinformatics that resides in large-scale sequencing centers. This hybrid model could be implemented in several ways, such as encouraging individual investigators to establish collaborations with scientists from large-scale centers or having funding agencies facilitate these interactions. Given the wide range of organisms that are being considered for genome analysis, it is likely that multiple models of collaboration will be needed.

Design of such a virtual approach to providing sequencing facilities would take careful thought. It would need to ensure that a diverse portfolio of activities, occurring at different locations and involving different teams of researchers, were adequately coordinated. It would need some mechanisms to minimize duplication of effort and identify potential opportunities for cost-sharing among partners. Some lessons could be learned from experiences gained with the University-National Oceanographic Laboratory Systems (UNOLS), which has shown how academic institutions and national laboratories can join forces and facilitate coordination among researchers with a common equipment need (e.g., ships). There are also lessons to be learned from the Ocean Drilling Program (ODP) (now the Integrated Ocean Drilling Program or IODP) experience, a long-term international research program that links universities, industry, and government.

As described in Chapter 3, coordination in the selection of organisms for sequence analysis will be critical. This is particularly important for microorganisms, given that sequencing is being carried out in a variety of ways, in laboratories around the world, and with funding from a number of federal and private agencies. In the United States, an Interagency Working Group on Microbial Genomics has been established by the National Science and Technology Council to provide a forum for discussion among interested parties about priorities and objectives and to build a more efficient coalition in a coordinated manner. Although prioritization of organisms for sequencing is difficult, the criteria outlined in Chapter 3 provide an initial framework and a starting point for further discussion. As costs for DNA sequencing continue to decrease, the number of possible species that can be selected for sequence analysis will continue to grow. The polar research community needs to provide input into the selection discussion through specific workshops to discuss these questions and through individual investigator-initiated proposals. At this point in the development of genomics, any additional genome sequencing projects related to polar organisms will add considerably to the existing body of knowledge.

## Sample Repositories and Culturing Facilities

Establishment of a base-funded and staffed repository for frozen samples of polar organisms would serve functions that are becoming increasingly important as new genomic and postgenomic technologies are being trained on problems of polar biology. Such a facility would:

- provide long-term maintenance and curation of frozen samples, thereby allowing genetic comparisons with natural populations in the future;
- offer new investigators in polar biology the opportunity to obtain samples to conduct pilot studies that could ultimately form the basis for a more complete proposal without the need for actual deployment to field sites; and
- provide access to samples of polar organisms to the broader community of biologists.

The most probable site for such a facility would be associated with a university or other research institution. Sample storage would most probably be in a "freezer farm" of ultracold freezers, each equipped with a liquid-nitrogen cryogenic backup system. At least one full-time staff person (initially) would be needed to oversee curation and maintenance. Operation of such a facility would require:

- developing a system of labeling and inventory control for management and retrieval of samples and
- archiving and curation of samples submitted to the repository.

Ancillary requirements of infrastructure would include:

- establishment of a means of ensuring the quality and identity of samples submitted for deposition (e.g., how do we know that it is *really* from the species purported by the submitter? How can we be sure that the sample has been handled in a satisfactory fashion prior to deposition?);
- establishment of an oversight committee or manager to evaluate the scientific legitimacy of requests for samples; and
- development of a set of clear conditions and expectations for use of samples that the requesting investigator would have to accept before samples would be released from the repository.

After establishing such a facility, individual polar investigators could respond to other scientists' requests that samples be collected for them by directing them to the facility, with explanation that a codified process has been established for such a request.

## Arctic Biology Laboratories

Arctic marine and terrestrial biological research conducted by U.S. scientists is currently limited by the small number of U.S.-supported research facilities and their lack of sophisticated instrumentation (see Plate 7). This problem is ameliorated to some extent by international agreements that permit access by U.S. scientists to the excellent facilities maintained by Canada, the Fennoscandian countries (Denmark, Finland, Norway, and Sweden), and Greenland. Access to the research facilities of Russia is, at present, problematic.

The two major U.S. research facilities are located in the Arctic at Barrow (71.3°N, 156.78°W) on the Arctic coast of Alaska and Toolik Field Station (68.6°N, 149.6°W) in the foothills of the Brooks Range on the shore of Toolik Lake. The Barrow facility began as a Navy laboratory and is now owned and operated by the Ukpeagvik Inupiat Corporation (The Barrow Village Corporation). The Barrow Arctic Science Consortium provides logistics support for research projects. The facility provides access to marine, sea-ice, coastal, inland tundra, and freshwater ecosystems. Lodging is limited, and laboratory equipment is spartan (<www.sfos.uaf.edu/basc/>), but year-round science support is available. Toolik Field Station, established in 1975 and managed by the University of Alaska's Institute of Arctic Biology, supports research in lake, riparian, taiga, tundra, and wetland environments. The station can support 40 researchers under normal circumstances and up to 80 for short periods. Laboratory equipment on-site is modest, and there is limited support for molecular and genomic research.

Currently, only a handful of researchers conduct genomic research in the Arctic, and this may be a reflection of the absence of high-technology facilities. In the near term, the creation of a molecular biology laboratory will facilitate the application of new biological tools in Arctic bioscience. This laboratory should include capabilities such as real-time PCR (TaqMan) and a microarray reader for monitoring the regulation of gene transcription. (Production of gene chips and sequencing are best done elsewhere as described in Chapter 3.) These technologies would be useful to investigators irregardless of the length of their field seasons. For those with long periods on-site, seasonal processes could be monitored and data collected and analyzed in real time. Short-term investigations of particular events would also benefit because P.I.s could verify that the phenomenon of interest is occurring before commencing intensive sampling; currently, such sampling must be conducted blindly. Similar arguments support the provisioning of Arctic facilities with two-dimensional gel electrophoresis equipment for proteomic work. NSF is to be commended for its efforts to upgrade these facilities, including buildings, laboratory equipment, and science support personnel.

Polar biologists from the U.S. should be urged to access research sites in the Canadian high Arctic that provide unique habitats (e.g., the perennial salt springs on Axel Heiberg Island and other sites that are analogous to the permanently ice-covered lakes, glacial ice, and subglacial soils of the Antarctic). The Canadian Polar Continental Shelf Project (PCSP) coordinates support for Canadian government and university scientists and non-Canadian researchers working in isolated areas throughout the Canadian Arctic (<http://polar.nrcan.gc.ca/home_e.html>). Non-Canadian researchers are required to have secured funding prior to applying for logistical support by PCSP. PCSP support includes transportation, communication, accommodation, field equipment, and related services. PCSP research facilities do not have any laboratory instrumentation or high-technology equipment, so the conduct of genomic research there may be very limited.

U.S. biological research in the eastern Arctic has traditionally been supported through an international agreement with Denmark. Major facilities available to U.S. biologists include the Danish stations Kangerlussuaq (southwest coast of Greenland, 67.0°N, 50.7°W ) and Zackenberg (northeast coast; 74°30' N, 20°30' W) and the U.S. air base at Thule (northwest coast; 76.5°N, 68.8°W). Ecosystems available at these locations include coastal, marine, riparian, sea ice, tundra, and wetland. Given its extreme northern location and relative ease of access via weekly Air Mobility Command flights, 109th Air Wing flights, and commercial flights from Copenhagen and Kangerlussuaq, Thule can potentially be developed as another major marine-terrestrial research station.

Svalbard (also known as Spitsbergen) offers a unique opportunity for the development of an international polar marine-terrestrial biology station. The island, ~79°N, is located near the average permanent extent of northern sea ice and is home to the Ny-Alesund Large-Scale Facility for Arctic Environmental Research. The Atmospheric Climate Research and Biological Research Facilities of the Norwegian Polar Institute support research on physical parameters that affect climate, on ultraviolet (UV) radiation and its biological effects on marine and terrestrial ecosystems and on the marine ecology of Arctic glacial fjords. Ny-Alesund is open primarily to researchers from the European Community. NSF is negotiating an agreement to open the facility to U.S. polar scientists, and this access could have significant payoffs for polar scientists in general and biologists in particular. Access to the climate research center will allow complementary studies on the effects of global climate change and UV radiation across the Arctic and in the Antarctic. Furthermore, comparative genome analyses of organisms in both poles can elucidate polar organisms' strategies for acclimation to the changing environment.

### Year-round Access to Polar Facilities

Winter at high latitudes poses exposure to one of the most extreme low-temperature, aphotic environments on Earth. Polar ecosystems are end-members of significant global importance. They represent a natural laboratory in which unique adaptations can be elucidated, and their origin and evolution understood, because most polar organisms are not just "surviving the extremes" but are actively feeding, growing, and reproducing. Study of polar ecosystems in winter will yield new information that can be used to identify and begin to understand the physiological processes and evolutionary pathways that lead to adaptation to life under extreme conditions.

Most scientific activities at high latitudes are currently limited to the sunlit spring, summer, and autumn periods. Access to Arctic and Antarctic terrestrial field sites is severely limited by logistics and safety concerns in the winter. In the marine environment, access by traditional surface vessels to polar waters is limited due to the presence of pack ice, especially in winter, and is hazardous because the waters are often poorly charted and can contain icebergs. Although seasonal constraints on scientific research may not be a significant impediment for some studies, for others the lack of access and physical presence means that datasets are incomplete. All liquid water systems (for example, in and beneath sea ice, lakes and brine ponds, subglacial lakes) in polar regions support life. The life-supporting behavioral and biogeochemical processes occur year-round; these processes do not cease during winter darkness. Consequently, knowledge and understanding of the seasonal variability of physical and biological processes and interactions in these systems are severely limited by a lack of winter data collection, direct observation, and experimentation.

Annual datasets are necessary for the following reasons:

- Meteorological conditions control biogeochemical rates and fluxes.
- Without coverage of the annual cycle, physical, chemical, and biological balances cannot be constructed for temporal comparisons with other global systems.
- Overwintering strategies of all life forms are crucial for understanding the persistence and evolution of organisms in these climates.

Their life history strategies cannot be deduced from studies in other habitats and ecosystems or from the summer season only.
• Processes that occur during winter (at any latitude) are inextricably linked to summer processes and vice versa. NSF's Long-Term Ecological Research (LTER) initiative realized the significance of obtaining data on annual scales and stressed the importance of annual material balances for cross-site comparisons and for assessing long-term data trends.
• Annual datasets allow an assessment of immediate ecosystem response to global change and provide information to understand how biodiversity and biocomplexity are related to global changes.

Some ideas and potential solutions that address constraints on winter access to polar regions are as follows:

• Initiate transition from "daylight-only" to year-round access for field stations such as Toolik Lake and McMurdo. Longer access will also allow greater flexibility for a broad range of scientists to participate directly in field research and will encourage new participants to enter polar research. This new thinking should, in turn, produce new insights on polar ecosystems, enhanced scientific interaction, and synergy of ideas.
• Request access to a declassified nuclear submarine with no or minimal military missions. The Science Ice Expeditions (SCICEX) program organized jointly by the Office of Naval Research and NSF demonstrated that nuclear submarines are effective sampling platforms for frozen oceans and not just for marine geology and geophysics or physical oceanographic measurements (Langseth et al., 1993). The SCICEX program generated unprecedented data on the biogeography of Arctic microbial populations (Bano and Hollibaugh, 2000, 2002; Ferrari and Hollibaugh, 1999) and also hosted studies of macrobiota such as amphipods. Although this program has ended, such programs could benefit polar genomics-enabled research by enabling survey work unimpeded by ice and access to winter populations of marine organisms. Further, this platform is conducive for the deployment of autonomous samplers.
• Establish floating observatories such as the one used in the joint U.S.-Canada Surface Heat Budget of the Arctic Ocean (SHEBA) program (Levi, 1998) and the one used in the international program led by Canada known as Canadian Arctic Shelf Exchange Study (CASES) currently planned for 2003-2004. The SHEBA program was modeled after the ice camps that have been used in Arctic exploration for more than a century but featured a floating marine station in the form of a Canadian icebreaker that was allowed to freeze into the ice. The camp drifted with the ice pack for one year and 800 km, providing opportunities for sampling a

host of biological and physical variables. The SHEBA program and its predecessors such as the Trans Arctic cruise of 1994 (Wheeler, 1997, and other papers in that volume) also demonstrate the crucial role that international collaborations can play in U.S. polar research. Neither of these programs would have been possible without collaboration with Canadian scientists.

• Continued development of autonomously and remotely operated vehicles (AOVs and ROVs) and moorings may provide mechanisms for obtaining samples relevant to genomics-enabled research. Possible applications, besides sample collection, may be the *in situ* detection of specific organisms that would permit time series analysis of population dynamics or studies of vertical or horizontal distributions that are currently constrained by ice cover.

**Coastal Research Vessels**

Polar marine biology is ultimately dependent on the availability of a fleet of vessels with complementary mission capabilities. In the Arctic, the U.S. Coast Guard Cutter *Healy* (WAGB 20, 420 ft.) and the UNOLS R/V *Alpha Helix* (133 ft.) provide research platforms of large and intermediate-small size for multiseason research at high northern latitudes. In the near future, the *Alpha Helix* will be replaced by the larger, as yet unnamed, Arctic Region Research Vessel (226 ft. overall). Other large and intermediate vessels of the UNOLS fleet frequently work in boreal polar waters (and occasionally in the Southern Ocean). Near-shore research is supported by smaller vessels such as the R/V *Montague* (58 ft.) of the Alaska Department of Fish and Game and the R/V *Tiglax* (~120 ft.) of the U.S. Fish and Wildlife Service. Beyond these U.S. resources, the vessels of other nations that border the Arctic are sometimes available on an ad hoc basis. Development of jointly funded projects can enable unique, highly productive expeditions, such as the U.S.-Canada SHEBA program (Levi, 1998) and the International North Water Polyna Study (Deming et al., 2002). International cooperation with Canada and other nations enhances access to the North American Arctic coastline and the central Arctic basin.

The history of the U.S. Antarctic research vessel fleet is one of growth in size and capability, yet of limitation in number. Various U.S. Coast Guard icebreakers (for example, the *Wind* class, the *Glacier*, the *Polar* class) have provided some research support, but their major function has been to open the channel to McMurdo Station in support of the annual logistics resupply. Mission-oriented research vessels have included the R/V *Hero* (a 130-ft., motor-sailor that primarily served Palmer Station from 1968 to 1985); the R/V *Polar Duke* (a 219-ft. ice-strengthened research and supply vessel with substantial oceanographic capability, in service 1985-1997, largely in the Antarctic Peninsula-Palmer Station region); the R/V *Nathaniel B. Palmer* (a 308-ft. ice-breaking, multidisciplinary research vessel that entered service in 1992); and the ARSV *Laurence M. Gould* (230-ft., the *Polar Duke's* replacement, in service 1997 to the present).

The design and construction of the *L.M. Gould* has provided a further significant advance in the sophistication of shipboard science. Demand for use of the *Gould* in support of deep water and more distant offshore science has been increasing steadily since her deployment, inevitably leading to heightened competition with demands from near-shore science based at Palmer Station. Furthermore, near-shore work is constrained by the *Gould's* size and draft. To enhance science opportunities in the Antarctic Peninsula while permitting the *Gould* to be exploited more effectively as an oceanographic research platform, acquisition or charter of a third, smaller research vessel is under discussion.

## U.S. National Ice Coring Laboratory

The U.S. National Ice Core Laboratory (NICL) is a facility for storing, curating, and studying ice cores recovered from the polar regions of the world. It provides scientists with the capability to conduct examinations and measurements of ice cores, and it preserves the integrity of these ice cores in a long-term repository for current and future investigations (<http://nicl.usgs.gov/index.html>). With the discovery of numerous subglacial lakes in Antarctica and related discoveries on the biodiversity, evolution, and physiology of the organisms present in deep ice cores, enhancing and expanding the current facility will be necessary to accommodate additional deep ice core samples for biological analysis.

## Rationalizing and Improving Specimen Transport

Given the great importance and high logistical costs of obtaining scientific samples in polar regions, it is imperative that samples are handled and returned to home institutions carefully. Samples are returned from Antarctica to the United States either by vessel shipment to California followed by surface or domestic airfreight to home institutions or via military and commercial air shipment directly from Antarctica either as hand-carried or checked baggage. This system generally works well, but on occasion valuable samples have warmed unacceptably during transit. Figure 4-1 presents data from a temperature logger placed with samples during vessel shipment from McMurdo Station to Point Hueneme, California, showing that samples warmed from near –20°C to 0°C and cooled

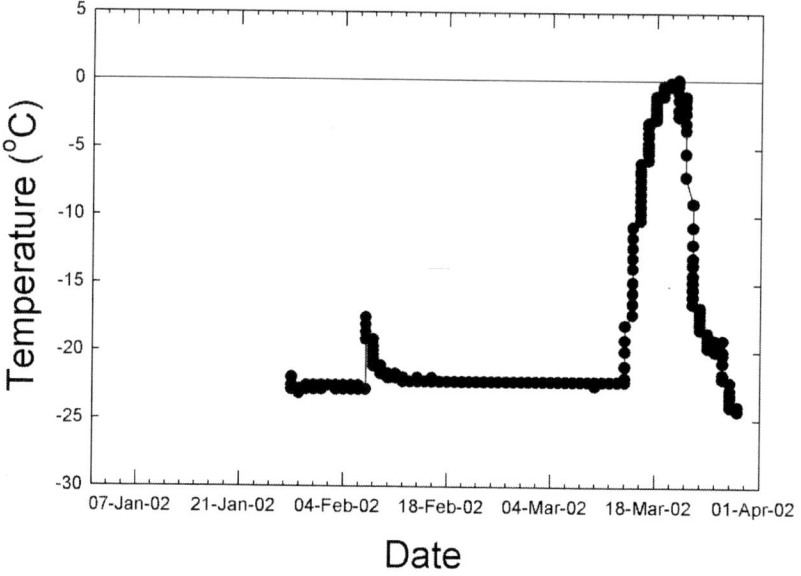

FIGURE 4-1 Sample temperature recorded during shipment of samples from McMurdo Station to Point Heuneme, California. Note that the samples reached 0°C for a 12-day period during shipment; the requested shipping temperature was –20°C. SOURCE: J.C. Priscu.

back to –20°C over a 12-day period. Such warming can compromise many types of samples. Without the temperature record made during shipment, scientists could misinterpret data subsequently obtained from compromised samples. Care must be taken to maintain desired temperatures during shipment, and all investigators should be presented with detailed temperature records during the period of shipment.

# 5

# An Integrated Polar Biology Community: Interactions Among Scientists, Education, and Outreach

## FACILITATING INTERACTIONS AND TECHNOLOGY TRANSFER ACROSS SCIENTIFIC DISCIPLINES

To develop a robust theoretical and empirical understanding of organisms and their roles in polar ecosystems requires an integration and synthesis of knowledge gained in many fields, from the biology of the organism to the physical and chemical characteristics of its environment. The Arctic and Antarctic polar science communities now have unique new opportunities to use multidisciplinary research and an array of new technologies to address questions that seemed unanswerable just a decade ago. However, success will require collaboration and interchange of information. Collaborations—whether interdisciplinary, national, or international—are typically more difficult for polar researchers than for scientists working in other regions. This chapter explores some of the impediments to collaborative efforts and possible avenues for improving collaboration.

### Building an Integrated Polar Community

Recent reports have addressed the urgency and complexity of global and environmental problems (NRC, 1999, 2001; NSB, 2000; NSF ACERE, 2003; PCAST, 1998). The reports acknowledged that many scientific disciplines are required to understand the interacting and interdependent components of the biosphere and Earth system science. The advancement of many disciplines and the contribution of new technologies—including communication technologies—to the social, natural and physical sciences are also noted in those reports. Most importantly, the exchange and integration of knowledge within and across environmental disciplines have been given high priority in all these reports.

The challenge of integrated research for the polar scientific communities is considerable, because many unique factors contribute to a separation within and across the Arctic and Antarctic scientific communities. At the simplest level is the difference in research field seasons. Most Antarctic work is conducted during the austral spring and summer (October to February), and most Arctic work during the boreal spring, summer and fall (March to November). Thus, many polar biologists must rely on collaborations with others to accomplish comparative studies of similar habitats in the Arctic and Antarctic. A higher order problem is the lack of a scientific society dedicated to polar biology and related disciplines. Such a society would raise the profile of all polar research areas while providing a forum for researchers to establish collaborative, bipolar research programs. Examples of focused biological organizations that provide highly interactive and integrative research venues are the self-organized *Drosophila*, worm (*Caenorhabditis elegans*), and zebrafish communities (see representative web sites <http://flybase.bio.indiana.edu/>, <http://biosci.umn.edu/CGC/CGChomepage.htm>, and <http://zfin.org/>, respectively). Third, the National Science Foundation's Office of Polar Programs (NSF OPP) administers and manages research for the Arctic and Antarctic separately. Although this approach is logical, it presents challenges to research groups that attempt to conduct integrated bipolar activities. Fourth, proposals that include international components (collaborators or facilities) face challenges due to NSF policies and considerations in the partner nations as well. International research collaborations in the Antarctic, for example, may require negotiation of a memorandum of understanding or other agreement with another nation's research program, which increases the bureaucratic burden for scientists as well as NSF. Logistic impediments (e.g., funding for stipends and travel, visas, unavailability of ship or aircraft time) also contribute to the lack of interchange between scientists from different nations. Perhaps NSF could solicit comments from the polar science community through a survey that would identify major impediments to bipolar research and international collaboration. Strategic implementation of cogent suggestions should help to solve some of these problems.

Enhancing our understanding of organismal, ecological, and ecosystem science through the use of techniques such as genome science will benefit from communication with other areas of biology and Earth system science. Biophysical and biogeochemical coupling addresses questions such as the following:

- What is the set of natural resources necessary to maintain life?
- How are organisms common to both polar regions expected to respond to nonlinearities of the hydrologic cycle predicted under climate change?
- How does snowfall distribution affect the spatial pattern of plant distribution across polar latitudes?
- Does the biodiversity in hot and cold extreme habitats have similar survival (gene-based) strategies?

Clearly, building a mechanism to encourage information exchange and collaboration across a community of scientists within habitats (oceans, soils, aquatic systems, ice), across habitats within each polar region, and across both polar regions should be a high priority. Enhancing data integration, syntheses, and knowledge in turn presents more opportunities for polar studies to be seen as integral for comparisons to other ecosystems and the biosphere.

### Example of an Integrative Program

NSF OPP's Arctic System Science (ARCSS) Program has made admirable strides in uniting the Arctic scientific community in a relatively short time, both within the United States and internationally. The ARCSS program was designed to advance the scientific basis for predicting environmental change and for formulating policy options in response to the anticipated impacts of global change on humans and societal support systems (see web site: <http://www.nsf.gov/od/opp/arctic/system.htm>). To achieve its goal, ARCSS promotes the understanding of physical, geological, chemical, biological, and sociocultural processes of the Arctic system. ARCSS has been successful in uniting polar scientists from various disciplines by supporting large integrated research projects that are proposed and implemented in response to science plans developed by the scientific community through Science Steering Committees. Furthermore, ARCSS has been particularly good at using web-based and e-mail communications to broaden participation and the sense of community. Lessons from that program could benefit polar biology, especially should there be a genome initiative.

### Working Groups and Workshops

Strengthening interactions within the polar community can be accelerated by providing new opportunities for small amounts of funding for scientific workshops and working groups. Workshops and working groups should consider the basis for understanding the fundamental processes of biology through synthesis of concepts, theory, comparisons, and contrasts. This approach allows teams of scientists that work on the same or many different organisms (thus, many disciplines) within a particular habitat (lakes in the Antarctic, soils in the Arctic, ice shelves), as well as between poles (lakes in the Arctic and Antarctic, and so on), to share information from various techniques and to develop new insights.

The National Center for Ecological Analysis and Synthesis (NCEAS) funded by NSF, is illustrative of how a small international working group can address a scientific topic and synthesize data and information. NCEAS has held a series of meetings over one to two years involving different participants. A key activity of these working groups is funding risky projects to address novel scientific questions and to support syntheses that might not be funded through traditional NSF channels. Easy access to data from many sources (bioinformatics for genomes, environmental data) is essential. Statisticians familiar with metadata and other statistical analyses, as well as scientists experienced in geographic information systems (GIS) and modeling are frequently integral to the workshops. A recent NSF review (NSF, 2002) highlighted the success of NCEAS working groups for the advancement of new theory and concepts, integration and synthesis of science, and facilitating communications across disciplines. Another example is the International Arctic Polyna Programme (IAPP) created by the Arctic Ocean Studies Board (AOSB) to address the physical and biological role of three polynas in the Arctic. AOSB charged a small group of international scientists (the Science Coordination Group) to define the scientific needs and to coordinate the execution of research.

### International, Multidisciplinary, Integrative Funding Initiatives

As genomics technologies begin to be applied to the study of physiological mechanisms of polar organisms and their response to physical stress, the need for international multidisciplinary research will likely emerge. The sequencing and analysis of the model plant species *Arabidopsis thaliana*, carried out by the *Arabidopsis* Genome Initiative comprised of scientists from large and small laboratories in the United States, Great Britain, France, and Japan, represented a successful model for collaboration among international scientists with expertise in genomics, bioinformatics, and plant biology. Although the six research groups secured major funding from agencies in the participating countries (from NSF, the Department of Energy (DOE), and the Department of Agriculture in the U.S.), they are part of a single project. Representatives of the six groups met to discuss strategies for facilitating international cooperation in completing the genome project and to establish a memorandum of understanding. Two key factors allowed for completion of this project several years ahead of schedule. The first was the willingness of the participating laboratories to work as a team to ensure that the project proceeded as quickly as possible. In many cases this meant that original work assignments were revised so that all laboratories were operating at maximum capacity throughout the project. Second, the distributed workload meant that the costs of the project were shared by funding agencies within the participating countries. Development of an integrated, international, and multidisciplinary polar genome initiative is likely to require cross-directorate funding within the NSF as well as funding by other agencies. At NSF, integrative biology and genomic research is funded by the Directorate of Biological Sciences, whereas polar research and logistical support are primarily funded by the Office of Polar Programs. Other funding agencies that support genomic research (for example, DOE's Genomes to Life program and the National Institutes of Health's National Human Genome Research Institute) do not provide logistical support for polar research. Thus, administrative coordination across NSF directorates and among funding agencies will be essential to facilitate integrative genomic research in polar regions.

Existing multidisciplinary research programs can also be complemented by the polar genome science initiative described in Chapter 3. NSF has had great success with the Long-Term Ecological Research (LTER) Network. The network is a collaborative effort involving more than 1,100 national and foreign scientists and students. It promotes synthesis and comparative research across sites and ecosystems and among other related national and international research programs over long temporal scales. Along with the other 21 LTER sites, research at the Alaska, Palmer, and McMurdo Dry Valleys LTER sites has addressed a range of questions from genomics to ecosystem-level science. The long-term environmental datasets from the LTERs and information gathered by new genomic technologies allow scientists to determine how organisms may change, adapt, and evolve in response to the changing environment. LTERs also allow comparative genomic studies on organisms in comparable conditions at both poles that could answer a number of the research questions outlined in Chapter 2.

## EDUCATION AND OUTREACH

There are several compelling reasons why the flow of information about polar biology to a wider audience should be enhanced. Perhaps the most important is the significant role that polar ecosystems play in global-scale phenomena. Polar organisms, while fascinating examples of adaptation to environmental extremes, also have a strong bearing on understanding ecological systems at lower latitudes. In terms of the effects of global climate change, notably rising temperatures and changes in ultraviolet (UV) radiation, polar organisms may prove to be "canaries in the coal mine" that provide early indications of how change might be affecting ecosystems. For example, because of their sensitivity to rising temperatures, polar organisms could offer an early glimpse into phenomena that may occur in ecosystems throughout the world. The diminishing health of polar organisms and ecosystems is already impacting the daily lives and health of the indigenous people of the Arctic. Therefore, it is important to learn more about polar biology and to communicate what is being discovered widely and rapidly.

Efforts to educate the public about polar science could be targeted at a wide range of lay and scientific audiences. The key mechanism for reaching nonscientists is the mass media. To reach young people, science texts used in secondary and university-level education have to include information about polar organisms and ecosystems. They should convey both the excitement of the polar environment and the relevance of the polar regions to pressing questions, ranging from insights into how biomolecules like proteins work, how new types of adaptive traits arise, and how global climate change disrupts the functioning of individual organisms and ecosystems as a whole. This same sense of excitement and challenge has to be conveyed to the research community in order to attract to polar science the types of expertise needed and a next generation of creative minds.

There are three general target audiences that could be reached through educational and outreach efforts: (1) K-12 and college education, (2) the research community, and (3) local communities in the Arctic region.

### K-12 and College Students

- *Expand coverage of polar topics in textbooks* (or develop textbooks that focus on polar science, broadly defined). Although a text devoted to polar biology might not be suitable for introductory-level classes, upper-division undergraduate classes and graduate seminars might be excellent contexts for presenting the information in such a book or monograph.
- *Use modern educational technology such as real-time distance learning to bring students into close contact with polar biology.* Technology exists for transmitting in real time (or recording for later presentation) the field studies being carried out by polar biologists. An example of how this technology might work is provided by the real-time transmission of activities on oceanographic vessels, for example, the activities of manned submersibles and remotely operated vehicles (ROVs). The Monterey Bay Aquarium regularly projects real-time images of ROV activities being conducted by its sister institution, the Monterey Bay Aquarium Research

Institute (MBARI). Live contacts with the scientists participating in these expeditions have proven to be an exciting and educationally successful mode of communicating science.

- *Develop strategies for bringing teachers and students into the field.* Field expeditions to the Arctic and Antarctic for teachers and students allow them to see first-hand what polar organisms are like, how they interact, and how they are studied. The Teachers Experiencing Antarctica and the Arctic (TEA) program for K-12 teachers is one example of this type of program (see web site <http://tea.rich.edu/>)
- *Web Sites.* Web sites could provide attractive, informative, and up-to-date exposure of new audiences to polar biology. A successful model is the web site for the National Oceanic and Atmospheric Administration's Ocean Exploration program, which featured real-time images and "log" updates by scientists during the first biology exploration of the deep Canada Basin in September 2002. Provision of curricular materials that can be downloaded from a web site could improve the instructional value of polar biology.
- *Polar scientists should be proactive in communicating their discoveries to the media.* If the media are to present increased coverage of polar biology, then polar biologists must fully engage existing university and other institutional resources for contacting the media and explaining why their work merits press, radio, and TV coverage or assume that burden directly. Polar scientists should take advantage of NSF's program on "Communicating Research to Public Audiences," which provides funds for scientists to disseminate research results, research in progress, and research methods to public audiences through media presentations, exhibits, or youth-based activities (see web site <http://www.nsf.gov/pubs/2003/nsf03509/nsf03509.html>). Training and media programs such as the Aldo Leopold Leadership Program also teach academic environmental scientists to communicate to the public and media effectively and offer another means of training scientists and their graduate students to interact with the media.

## The Research Community

- *Develop field courses to encourage doctoral students, postdoctoral-level scientists, and established investigators to work on polar biology issues.* The Antarctic integrated biology course that was offered for six years in the Crary Laboratory at McMurdo Station might serve as an appropriate model. One of the many strong points of a course of this type is that the participants comprised a wide range of intellectual interests, such that disciplinary boundaries were surmounted very effectively. It is noteworthy that several of the participants in this course have continued to work on polar organisms, in several cases with federal grant support that has been awarded for their new lines of study.

- *Develop workshops to educate scientists about prospects for working in polar regions.* Many scientists are apt to regard work in polar regions as logistically complicated, inconvenient in terms of time requirements, and "not worth the effort" involved. To provide an accurate portrayal of what is involved in doing research in polar regions, workshops should be held to familiarize potential polar investigators with the requirements and opportunities for polar work. Included in these workshops should be a statement of the opportunities that exist for winter season work at polar laboratories, notably at McMurdo Station, where excellent research facilities generally lie idle during the winter season.
- *Develop small, focused meetings on polar biology modeled after the Gordon Research Conference (GRC) format.* Although symposia on polar topics often occur within larger meetings, relatively small meetings focusing on polar biology that are attended by scientists at different career stages and from many countries could be an excellent vehicle for educating polar biologists about the activities of their peers. Two examples to build upon are the GRC on Polar Marine Sciences held biannually and a recent symposium supported by the Nordic Arctic Research program that gathered 20 Nordic graduate students with 20 pan-Arctic senior scientists for an educational retreat on Arctic ecosystems in Sigulda, Latvia.
- *Bring potential collaborators to field sites.* Principal investigators who are currently conducting polar field research should be encouraged and assisted to bring collaborators to the field sites. At a minimum, such visits would improve the collaborators' understanding of the biology under study. Such visits might also lead to increased involvement of the collaborators (or their students) in field work.
- *Offer support for postdoctoral fellows.* A federally funded fellowship program for postdoctoral researchers could facilitate the entry of new investigators into polar research. This program is currently under design at the National Science Foundation.
- *Offer support to new investigators.* NSF has an existing program that offers support for new investigators in polar science.

### Local Communities in the Arctic Region

- *Serve, engage, and respect indigenous communities.* In the case of Arctic science, educational and outreach activities must target the indigenous communities that are part of the ecosystem that is being studied. This goal can be reached via specially funded programs for teachers and students in the north, but it can also be facilitated through regular research awards. When a federally-supported research team enters the Arctic to undertake research, one of their responsibilities should include a visit to the local community leaders, and schools where possible, to explain the goals of the funded research and engage in dialogue about the science and methodologies involved. For example, when icebreakers enter coastal waters, arrangements could be made for local residents, especially school children, to visit the ship for an afternoon and witness the ongoing work. This approach has been taken in recent years by the Canadian Coast Guard icebreakers (to the mutual benefits of villagers and scientists) when a research project brings the ship near a local village. For smaller scale projects, individual research budgets could include funds to return to the local community at the end of the study and explain in person how the results may be of interest or importance to local residents.

- *Encouraging local communities to contribute to research activities.* This seems a sound approach for communicating what science is being conducted and why, and for facilitating the research. For example, engaging knowledgeable residents or talented students in long-term monitoring efforts may prove essential to evaluating the effects of environmental changes on various aspects of arctic biology. In some cases, the research may target local residents themselves as members of the ecosystem under study. As the environmental side of polar genomics develops more fully—as we are better positioned to evaluate metabolic expressions *in situ* with genetic information and tools—key issues will also include human residents as top predators. To the extent that local communities continue to depend upon local food sources for their sustenance, the bioaccumulation of contaminants will remain a serious human health problem. If climate warming brings new pathogens to the region, microbial genomics will become a critical tool for charting potential solutions. Partnerships between polar genomic sciences and the health and social sciences will increase.

- *Respecting local culture and customs of indigenous communities.* Asking local communities for input about research questions and incorporating native knowledge of polar biology can bring surprising rewards scientifically and socially, as documented by Krupnick and Jolly (2002). Respecting local culture and customs can open the door to sharing scientists' excitement about local biological issues among secondary school children, which may in turn facilitate entry of some of these students into research or related careers where they currently are underrepresented. Sensitivity to local knowledge and customs of indigenous communities may be a prerequisite to orchestrating some of the long-term monitoring programs discussed earlier and is certainly essential to conducting research that includes the resident as a member of the Arctic ecosystem under study.

# 6

# Findings and Recommendations

The new era of genomics is opening doors to an unparalleled realm of research questions, and polar scientists are poised to make great advances. This chapter discusses how to move forward on those questions and focuses on two issues: the important roles that genomic technology (and related enabling technologies) stand to play in polar research and the appropriate means for making these approaches available for a diverse community of research scientists. The recommendations span a wide range of considerations, including mechanisms for providing the necessary financial and logistical support for polar research, strategies for identifying optimal study systems, means for improved integration of research programs in Arctic and Antarctic regions, and improvements in the infrastructure of education, to broaden awareness of the importance of polar biology and, ultimately, increase participation in polar research. The challenges posed by the recommended actions are substantial. Yet, if these challenges are met successfully, polar biological research will attain a new level of sophistication and enhance its ability to make vital contributions to a wide range of disciplines, ranging from biomedicine to climate change.

## A NEW UNIFYING APPROACH TO POLAR BIOLOGICAL RESEARCH

**Finding 1:** Genome science is an addition to, not a replacement for, other approaches to the study of polar biology. The applica-

FINDINGS AND RECOMMENDATIONS                                    129

tion of new genomic technologies has the potential to be a unifying paradigm for polar biological sciences. Key opportunities include the following:

- Polar organisms and communities offer unique opportunities to study evolution using genome sciences.
- The use of genomic methods will give insights into the effects of global change on polar biota and biogeochemistry.
- Genome sciences have vast potential for elucidating function in microbial communities.
- Polar genome sciences could make broad contributions to biomedicine and biotechnology (for example, cryopreservation, cryosurgery, and cold functioning enzymes).
- A polar genome research initiative will provide important new information on the evolution, physiology, and biochemistry of polar organisms. Such information not only enhances our understanding of how polar ecosystems function, but also helps our search for life in icy worlds.

**Recommendation 1-1:** The National Science Foundation (NSF) should develop a major new initiative in polar genome sciences that emphasizes collaborative multidisciplinary research and coordinates research efforts. The polar genome science initiative could facilitate genome analyses of polar organisms and support the relevant research on their physiology, biochemistry, ecosystem function, and biotechnological applications.

**Recommendation 1-2:** A new polar genome initiative should capitalize on data from existing Long-Term Ecological Research and Microbial Observatory sites to take advantage of the long-term datasets and the geographical distribution of these sites. Additional approaches may be taken so that research can be conducted at sites with comparable conditions at both poles. For example, there is currently no marine site in the Arctic.

## COORDINATION IS ESSENTIAL

**Finding 2:** To facilitate the advancement of polar genome sciences, coordination of research efforts will be required to ensure efficient transfer of technologies, provide guidance to researchers on choosing organisms for genome analyses, and help in the development of new scientific initiatives. Coordination of research efforts should begin with syntheses of the available information, thereby avoiding duplication of research efforts. It should facilitate increased communication among the polar scientists and also with nonpolar scientists who have expertise in genomics and other technological advances applicable to polar studies.

**Recommendation 2:** NSF should form a scientific standing committee to establish priorities and coordinate large-scale efforts for genome-enabled polar science (for example, genome sequencing, transcriptome analysis, and coordinated bioinformatics databases).

## VIRTUAL GENOME SCIENCE CENTERS

**Finding 3:** Genomic technologies, both those currently available and those anticipated in the future, are applicable to some of the key questions in polar biology. However, the technical demands of genome science often transcend the resources of any individual researcher.

**Recommendation 3:** NSF should support some mechanism to facilitate gene sequencing and related genomic activities beyond the budget of any individual principal investigator, such as virtual genome science centers. The purpose of the virtual centers would be to provide infrastructure for individual researchers and to facilitate technology transfer among researchers. New infrastructure is not needed, rather some type of coordinating body (e.g., University National Oceanographic Laboratory System, Ocean Drilling Program).

## ENABLING TECHNOLOGIES

**Finding 4:** Enabling technologies are critical to the successful application of genomic technologies to polar studies.

**Recommendation 4:** Ancillary technologies such as observatories, ice drilling, remote sensing, mooring and autonomous sensors, and isotope approaches should be developed to support application of genomic technologies to polar studies.

## INCREASING AWARENESS AND EDUCATION

**Finding 5:** Polar systems play important roles in global-scale phenomena, and there is a need for enhanced flow of information about polar biology to a wide audience of scientists, policymakers, and the general public.

**Recommendation 5:** NSF should continue its efforts to make information about polar regions available to teachers, schools, and the public. Short- and long-term plans should be developed for increasing public awareness of polar biology. In the near future, postdoctoral fellowships in polar biology could be set up to encourage young scientists to enter the field. Long-term plans should include continued efforts to incorporate polar biology in college and K-12 curricula.

## IMPEDIMENTS TO INTEGRATED POLAR SCIENCE

**Finding 6:** Impediments to conducting multidisciplinary integrated polar science exist, including administrative, fiscal, and infrastructure issues:

- Coordination among directorates within NSF and coordination among agencies are both essential for advancing polar biology.
- International collaborations are vital for all polar research. Current procedures make the involvement of international scientists in U.S. polar biological projects difficult.
- Attempts to conduct comparative research at both poles can be difficult. Although NSF's Office of Polar Programs supports research in both poles, grant applications for Arctic and Antarctic research have to be made to two separate NSF research programs. Research proposals often undergo two reviews, and scientists must prepare separate budgets for each proposal.
- Infrastructure for Arctic and Antarctic biology needs improvement. The conduct of molecular research in the polar regions requires specific infrastructure, and there is no high-technology equipment for such work in the Arctic. Development of ice-drilling and clean sampling technologies in the Antarctic will facilitate research in deep ice and subglacial lakes.

**Recommendation 6-1:** To reach the goal of getting excellent science done as efficiently as possible, NSF should remove impediments to cross-directorate funding. Because integrated polar science often requires interagency cooperation, NSF should lead by example and form partnerships with the National Aeronautics and Space Administration and others as relevant. Memoranda of understanding among directorates within NSF and among funding agencies are one mechanism to facilitate transfer of information and coordination of research.

**Recommendation 6-2:** Establishment of international research partnerships or memoranda of understanding will facilitate and enhance these collaborative efforts. Issues such as stipends, travel, visas, education, ship time, aircraft use and other logistical issues should be addressed in these memoranda to ensure successful operation of international collaborative polar research.

**Recommendation 6-3:** More information is needed to develop solutions to problems related to conducting bipolar research. NSF should conduct a brief survey of researchers and research groups who would potentially work in both poles to identify impediments and then take steps to address them.

**Recommendation 6-4:** To facilitate integrated, multidisciplinary biological research at both poles, NSF will have to improve biological laboratories and research vessels, and develop ice-drilling resources in the polar regions. Opportunities to allow year-round access to, and operation of, field sites should be pursued.

# References

Aagard, K., D. Darby, K. Falkner, G. Flato, J. Grebmeier, C. Measures, and J. Walsh. 1999. *Marine Science in the Arctic: A Strategy*. Fairbanks, AK: Arctic Research Consortium of the United States (ARCUS). 84 pp.

Aakra, A., J.B. Utaker, and I.F. Nes. 1999. RFLP of rRNA genes and sequencing of the 16S-23S rDNA intergenic spacer region of ammonia-oxidizing bacteria: A phylogenetic approach. *International Journal of Systematic Bacteriology* 49:123-130.

Aghajari, N., G. Feller, C., Gerday, and R. Haser. 1998. Structures of the psychrophilic *Alteromonas haloplanctis* alpha-amylase give insights into cold adaptation at a molecular level. *Structure* 6:1503-1516.

Aguilar, P.S., A.M. Hernandez-Arriaga, L.E. Cybulski, A.C. Erazo, and D. Mendoza. 2001. Molecular basis of thermosensing: a two-component signal transduction thermometer in *Bacillus subtilis*. *EMBO Journal* 20:1681-1691.

Aislabie, J., J. Foght, and D. Saul. 2000. Aromatic hydrocarbon-degrading bacteria from soil near Scott Base, Antarctica. *Polar Biology* 23:183-188.

Alley, R.B., and R.A. Bindschadler, (eds.). 2001. *The West Antarctic Ice Sheet: Behavior and Environment*. Washington, DC: American Geophysical Union.

Alley, R.B., C.A. Shuman, D.A. Meese, A.J. Gow., K.C. Taylor, K.M. Cuffey, J.J. Fitzpatrick, P.M. Grootes, G.A. Zielinski, M. Ram, G. Spinelli, and B. Elder. 1997. Visual-stratigraphic dating of the GISP2 ice core: Basis, reproducibility, and application. *Journal of Geophysical Research* 102(C12):367-388.

Anderson, J.G., D.W. Toohey, and W.H. Brune. 1991. Free radicals within the Antarctic vortex: The role of CFCs in Antarctic ozone loss. *Science* 251:39-46.

Aparicio, S., J. Chapman, E. Stupka, N. Putnam, J.-M. Chia, P. Dehal, A. Christoffels, S. Rash, S. Hoon, A. Smit, M.D. Sollewijn Gelpke, J. Roach, T. Oh, I.Y. Ho, M. Wong, C. Detter, F. Verhoef, P. Predki, A. Tay, S. Lucas, P. Richardson, S.F. Smith, M.S. Clark, Y.J.K. Edwards, N. Doggett, A. Zharkikh, S.V. Tavtigian, D. Pruss, M. Barnstead, C. Evans, H. Baden, J. Powell, G. Glusman, L. Rowen, L. Hood, Y.H. Tan, G. Elgar, T. Hawkins, B. Venkatesh, D. Rokhsar, S. Brenner. 2002. Whole-genome shotgun assembly and analysis of the genome of *Fugu rubripes*. *Science* 297:1301-1310.

Arrigo, K.R., G.R. DiTullio, R.B. Dunbar, D.H. Robinson, M. Van Woert, D.L. Worthen, and M.P. Lizotte. 2000. Phytoplankton taxonomic variability in nutrient utilization and primary production in the Ross Sea. *Journal of Geophysical Research* 105:8827-8846.

Azam, F. 1998. Microbial control of oceanic carbon flux: The plot thickens. *Science* 280:694-696.

Bae, W.H., B. Xia, M. Inouye, and K. Severinov. 2000. Escherichia coli CspA-family RNA chaperones are transcriptional antiterminators. *Proceedings of the National Academy of Science USA* 97:7784-7789.

Baker, C.S., G.M. Lento, F. Cipriano, M.L. Dalebout, and S.R. Palumbi. 2000a. Scientific whaling: Source of illegal products for market? *Science* 290:1695-1696.

Baker, C.S., G.M. Lento, F. Cipriano, and S.R. Palumbi. 2000b. Predicted decline of protected whales based on molecular genetic monitoring of Japanese and Korean markets. *Proceedings of the Royal Society of London B* 267:1191-1199.

Bakermans, C., and E.L. Madsen. 2002. Detection in coal tar waste-contaminated groundwater of mRNA transcripts related to naphthalene dioxygenase by fluorescent *in situ* hybridization with tyramide signal amplification. *Journal of Microbiological Methods* 50:75-84.

Bano, N., and J.T. Hollibaugh. 2000. Diversity and distribution of DNA sequences with affinity to ammonia-oxidizing bacteria of the β-subdivision of the class Proteobacteria in the Arctic Ocean. *Applied and Environmental Microbiology* 66:1960-1969.

Bano, N., and J.T. Hollibaugh. 2002. Phylogenetic composition of bacterioplankton assemblages from the Arctic Ocean. *Applied and Environmental Microbiology* 68:505-518.

Bargelloni, L., P.A. Ritchie, T. Patarnello, B. Battaglia, D.M. Lambert, and A. Meyer. 1994. Molecular evolution at subzero temperatures: Mitochondrial and nuclear phylogenies of fishes from Antarctica (suborder Notothenioidei), and the evolution of antifreeze glycopeptides. *Molecular Biology and Evolution* 11:854-863.

Barnes, B.M. 1989. Freeze avoidance in mammals: Body temperatures below 0°C in an arctic hibernator. *Science* 244:1593-1595.

Barrett, P.J., D.P. Elston, D.M. Harwood, B.C. McKelvey, and P.-N. Webb. 1987. Mid-Cenozoic record of glaciation and sea-level change on the margin of Victoria Land basin, Antarctica. *Geology* 15:634-637.

Becker, L.B., M.L. Weisfeldt, M.H. Weil, T. Budinger, J. Carrico, K. Kern, G. Nichol, I. Shechter, R. Traystman, C. Webb, H. Wiedemann, R. Wise, and G. Sopko. 2002. The PULSE initiative: Scientific priorities and strategic planning for resuscitation research and life saving therapies. *Circulation* 105:2562-2570.

Béjà, O., L. Aravind, E.V. Koonin, M.T. Suzuki, A. Hadd, L.P. Nguyen, S. Jovanovich, C.M. Gates, R.A. Feldman, J.L. Spudich, E.N. Spudich, and E.F. DeLong. 2000. Bacterial rhodopsin: Evidence for a new type of phototrophy in the sea. *Science* 289:1902-1906.

Béjà, O., E.N. Spudich, J.L. Spudich, M. Leclerc, and E.F. DeLong. 2001. Proteorhodopsin phototrophy in the ocean. *Nature* 411:786-789.

Béjà, O., E.V. Koonin, L. Aravind, L.T. Taylor, H. Seitz, J.L. Stein, D.C. Bensen, R.A. Feldman, R.V. Swanson, and E.F. DeLong. 2002a. Comparative genomic analysis of Archaeal genotypic variants in a single population and in two different oceanic provinces. *Applied and Environmental Microbiology* 68:335-345.

Béjà, O., M.T. Suzuki, J.F. Heidelberg, W.C. Nelson, C.M. Preston, T. Hamada, J.A. Eisen, C.M. Fraser, and E.F. DeLong. 2002b. Unsuspected diversity among marine aerobic anoxygenic phototrophs. *Nature* 415:630-633.

Behrenfeld, M.J., A.J. Bale, Z.S. Kolber, J. Aiken, and P.G. Falkowski. 1996. Confirmation of iron limitation of phytoplankton photosynthesis in the equatorial Pacific Ocean. *Nature* 383:508-511.

# REFERENCES

Benjamin, D.C., S. Kristjansdottir, and A. Gudmundsdottir. 2001. Increasing the thermal stability of euphauserase, a cold-active and multifunctional serine protease from Antarctic krill. *European Journal of Biochemistry* 268:127-131.

Bentahir, M., G. Feller, M. Aittaleb, J. Lamotte-Brasseur, T. Himri, J. P. Chessa, and C. Gerday. 2000. Structural, kinetic, and calorimetric characterization of the cold-active phosphoglycerate kinase from the Antarctic *Pseudomonas* sp. TACII18. *Journal of Biological Chemistry* 275:11147-11153.

Bergstrom, D.M., and S.L. Chown. 1999. Life at the front: history, ecology and change on Southern Ocean islands. *Trends in Ecology and Evolution* 14:472-477.

Berkman, P.A., and L.R. Tipton-Everett (eds.). 2001. *Latitudinal Ecosystem (LAT-ECO) Responses to Climate Across Victoria Land, Antarctica.* Report of a NSF workshop, BPRC Report No. 20. Columbus: Byrd Polar Research Center, Ohio State University Press. 152 pp.

Billings, W.D, J.O. Luken, D.A. Mortensen, and K.M. Peterson. 1982. Arctic tundra: A sink or source for atmospheric carbon dioxide in a changing environment? *Oecologia* 53:7-11.

Blank, C.E., S.L. Cady, and N.R. Pace. 2002. Microbial composition of near-boiling silica-depositing thermal springs throughout Yellowstone National Park. *Applied and Environmental Microbiology* 68:5123-5135.

Boetius, A., K. Ravenschlag, C.J. Schubert, D. Rickert, F. Widdel, A. Gieseke, R. Amann, B.B. Jørgensen, U. Witte, and O. Pfannkuche. 2000. A marine microbial consortium apparently mediating anaerobic oxidation of methane. *Nature* 407:623-626.

Borneman, J. 1999. Culture-independent identification of microorganisms that respond to specified stimuli. *Applied and Environmental Microbiology* 65:3398-3400.

Boschker, H.T.S., S.C. Nold, P. Wellsbury, D. Bos, W. de Graaf, R. Pel, R.J. Parkes, and T.E. Cappenberg. 1998. Direct linking of microbial populations to specific biogeochemical processes by C-13 labelling of biomarkers. *Nature* 392:801-805.

Both, C., and M.E. Visser. 2001. Adjustment to climate change is constrained by arrival date in a long-distance migrant bird. *Nature* 411:296-298.

Bowman, J.P., J.J. Gosink, S.A. McCammon, T.E. Lewis, D.S. Nichols, P.D. Nichols, J.H. Skerratt, J.T. Staley, and T.A. McMeekin. 1998. *Colwellia demingiae* sp. nov., *Colwellia hornerae* sp. nov., *Colwellia rossensis* sp. nov. and *Colwellia psychrotropica* sp. nov.: Psychrophilic Antarctic species with the ability to synthesize docosahexaenoic acid (22:6ω3). *International Journal of Systematic Bacteriology* 48:1171-1180.

Boyd, P.W., A.J. Watson, C.S. Law, E.R. Abraham, T. Trull, R. Murdoch, D.C.E. Bakker, A.R. Bowie, K.O. Buesseler, H. Chang, M. Charette, P. Croot, K. Downing, R. Frew, M. Gall, M. Hadfield, J. Hall, M. Harvey, G. Jameson, J. Laroche, M. Liddicoat, R. Ling, M.T. Maldonado, R.M. McKay, S. Nodder, S. Pickmere, R. Pridmore, S. Rintoul, K. Safi, P. Sutton, R. Strzepek, K. Tanneberger, S. Turner, A. Waite, and J. Zeldis. 2000. A mesoscale phytoplankton bloom in the polar Southern Ocean stimulated by iron fertilization. *Nature* 407:695-702.

Boyer, B.B., and B.M. Barnes. 1999. Molecular and metabolic aspects of mammalian hibernation. *Bioscience* 49:713-724.

Boynton, W.V., W.C. Feldman, S.W. Squyres, T.H. Prettyman, J. Bruckner, L.G. Evans, R.C. Reedy, R. Starr, J.R. Arnold, D.M. Drake, P.A.J. Englert, A.E. Metzger, I. Mitrofanov, J.I. Trombka, C. d'Uston, H. Wanke, O. Gasnault, D.K. Hamara, D.M. Janes, R.L. Marcialis, S. Maurice, I. Mikheeva, G.J. Taylor, R. Tokar, and C. Shinohara. 2002. Distribution of hydrogen in the near surface of Mars: Evidence for subsurface ice deposits. *Science* 297:81-85.

Braddock, J.F, M.L. Ruth, P.H. Catterall, J.L. Walworth, and K.A. Mccarthy. 1997. Enhancement and inhibition of microbial activity in hydrocarbon-contaminated arctic soils: Implications for nutrient-amended bioremediation. *Environmental Science and Technology* 31:2078-2084.

Brinton, K.L.F., A.I. Tsapin, D. Gilichinsky, and G.D. McDonald. 2002. Aspartic acid racemization and age-depth relationships for organic carbon in Siberian permafrost. *Astrobiology* 2:77-82.

Britten, R.J. 2002. Divergence between samples of chimpanzee and human DNA sequences is 5%, counting indels. *Proceedings of the National Academy of Sciences USA* 99:13633-13635.

Broughton, L.C. 2002. Understanding the role of soil microbial comunity in the Arctic tundra. P. 21 in Moran, M.A., and S.L. Cadee (eds.), *Microbial Observatories/LexEn Principal Investigators Workshop*. Arlington, Virginia: National Science Foundation.

Brown, J., K.R. Everett, P.J. Webber, S.F. MacLean, Jr., and D.F. Murray. 1980. The coastal tundra at Barrow. Pp. 1-29 in J. Brown, P.C. Miller, L.L. Tieszen and F.L. Bunnell (eds.). *An Arctic Ecosystem: The Coastal Tundra at Barrow, Alaska.* Stroudsburg, Penn.: Dowden Hutchinson, & Ross. 571 pp.

Browne, J., A. Tunnacliffe, and A. Burnell. 2002. Plant desiccation gene found in a nematode. *Nature* 416:38.

Bull, I.D., N.P. Parekh, G.H. Hall, P. Ineson, and R.P. Evershed. 2000. Detection and classification of atmospheric methane oxidizing bacteria in soil. *Nature* 405:175-178.

Buma, A.G.J., M.K. De Boer, and P. Boelen. 2001. Depth distributions of DNA damage in antarctic marine phyto- and bacterioplankton exposed to summertime UV radiation. *Journal of Phycology* 37:200-208.

Bunnell, F.L., D.E.N. Tait, P.W. Flanagan, and K. van Cleve. 1977. Microbial respiration and substrate weight loss. I. A general model of abiotic variables. *Soil Biology and Biochemistry* 9:33-40.

Calkins, J., and T. Thordardottir. 1980. The ecological significance of solar UV radiation on aquatic organisms. *Nature* 283:563-566.

Carpenter E.J., S. Kin, and D.G. Capone. 2000. Bacterial activity in South Pole snow. *Appliedand Environmental Microbiology* 66:4514-4517.

Castello, J.D., S.O. Rogers, W.T. Starmer, C.M. Catranis, L. Ma, G.D. Bachand, Y. Zhao, and J.E. Smith. 1999. Detection of tomato mosaic tobamovirus RNA in ancient glacial ice. *Polar Biology* 22:207-212.

Causton, H.C., B. Ren, S.S. Koh, C.T. Harbison, E. Kanin, E.G. Jennings, T.I. Lee, H.L. True, E.S. Lander, and R.A. Young. 2001. Remodeling of yeast genome expression in response to environmental changes. *Molecular and Cell Biology* 12:323-337.

Chapin, F.S., and G.R. Shaver. 1985. Individualistic growth response of tundra plant species to environmental manipulations in the field. *Ecology* 66:564-576.

Chen, T.H.H., and N. Murata. 2002. Enhancement of tolerance of aboitic stress by metabolic engineering of betaines and other compatible solutes. *Current Opinion in Biology* 5:250-257.

Chen, L., A.L. DeVries, and C.-H.C. Cheng. 1997a. Evolution of antifreeze glycoprotein gene from a trypsinogen gene in Antarctic notothenioid fish. *Proceedings of the National Academy of Sciences USA* 94:3811-3816.

Chen, L., A.L. DeVries, and C.-H.C. Cheng. 1997b. Convergent evolution of antifreeze glycoproteins in Antarctic nototheniid fish and Arctic cod. *Proceedings of the National Academy of Sciences USA* 94:3817-3822.

Chen, W.-J., C. Bonillo, and G. Lecointre. 1998. Phylogeny of the Channichthyidae (Notothenioidei, Teleostei) based on two mitochondrial genes. Pp. 287-298 in G. di Prisco, E. Pisano, E. and A. Clarke (eds.), *Fishes of Antarctica*. Milano: Springer-Verlag.

Chibber, B.A.K., P. Fay, G.H. Bernstein, F.J. Castellino, and J.G. Duman. 2002. Remote multisensing of biological systems. Pp. 883-887 in A.K. Hyder. et al., (eds.), *Multisensor Fusion*. Amsterdam: Kluwer Academic Publishers.

Christner, B.C., E. Mosley-Thompson, L.G. Thompson, and J.N. Reeve. 2001. Isolation of bacteria and 16S rDNAs from Lake Vostok accretion ice. *Environmental Microbiology* 3:570-577.

Chun, J., A. Huq, and R.R. Colwell. 1999. Analysis of 16S-23S rRNA intergenic spacer regions of Vibrio cholerae and Vibrio mimicus. *Applied and Environmental Microbiology* 65:2202-2208.

Chyba, C.F. 2000. Energy for microbial life on Europa. *Nature* 403:381-382.

Chyba, C.F., and K.P. Hand. 2001. Life without photosynthesis. *Science* 292:2026-2027.

Chyba, C.F., and C.B. Phillips. 2001. Possible ecosystems and the search for life on Europa. *Proceedings of the National Academy of Sciences USA* 98:801-804.

Cipriano, F., and S.R. Palumbi. 1999. Genetic tracking of a protected whale. *Nature* 397:307-308.

Clark, P. U., R.B. Alley, and D. Pollard. 1999. Northern hemisphere ice-sheet influences on global climate change. *Science* 286:1104-1111.

Clarke, A. 1990. Temperature and evolution: Southern Ocean cooling and the Antarctic marine fauna. Pp. 9-22 in K.R. Kerry, and G. Hempel (eds.), *Antarctic Ecosystems: Ecological Change and Conservation*. Springer-Verlag, Berlin and Heidelberg.

Clarke, A. 1991. What is cold adaptation and how should we measure it? *American Zoologist* 31:81-92.

Clarke, A., and I.A. Johnston. 1996. Evolution and adaptive radiation of Antarctic fishes. *Trends in Ecology and Evolution* 11:212-218.

Clifford, S.M., D. Crisp, D.A. Fisher, K.E. Herkenhoff, S.E. Smrekar, P.C. Thomas, D.D. Wynn-Williams, R.W. Zurek, J.R. Barnes, B.G. Bills, E.W. Blake, W.M. Calvin, J.M. Cameron, M.H. Carr, P.R. Christensen, B.C. Clark, G.D. Clow, J.A. Cutts, et al. 2000. The state and future of Mars polar science and exploration. *Icarus* 144:210-242.

Clow, G.D., and B. Koci. 2002. A fast mechanical-access drill for polar glaciology, paleoclimatology, geology, tectonics, and biology. *National Institute for Polar Research Special Issue* 56:1-30.

Cocca, E., M. Ratnayake-Lecamwasam, S.K. Parker, L. Camardella, M. Ciaramella, M., G. di Prisco, and H.W. Detrich, III. 1995. Genomic remnants of α-globin genes in the hemoglobinless Antarctic icefishes. *Proceedings in the National Academy of Sciences USA* 92:1817-1821.

Connon, S.A., and S.T. Giovannoni. 2002. High throughput methods for culturing microorganisms in very low nutrient media yield diverse new marine isolates. *Applied and Environmental Microbiology* 68(8):3878-85.

Cottrell, M.T., and D.L. Kirchman. 2000. Natural assemblages of marine proteobacteria and members of the *Cytophaga-Flavobacter* cluster consuming low- and high-molecular-weight dissolved organic matter. *Applied and Environmental Microbiology* 66:1692-1697.

Courtright, E.M., D.H. Wall, R.A. Virginia, J.T. Vida, L.M. Frisse, and W.K. Thomas. 2000. Nuclear and mitochondrial DNA sequence diversity in the Antarctic nematode *Scottnema lindsayae*. *Journal of Nematology* 32:143-153.

Crawford, M. 1987. Landmark ozone treaty negotiated. *Science* 237:1557.

Crockett, E.L., and B.D. Sidell. 1990. Some pathways of energy metabolism are cold adapted in Antarctic fishes. *Physiological Zoology* 63:472-488.

Cullen, J.J., and M.P. Lesser. 1991. Inhibition of photosynthesis by ultraviolet radiation as a function of dose and dosage rate: Results for a marine diatom. *Marine Biology* 111:183-190.

Cummings, C.J., and H.Y. Zoghbi. 2000. Fourteen and counting: Unraveling trinucleotide repeat diseases. *Human Molecular Genetics* 9:909-916.

D'Amico, S., C. Gerday, and G. Feller. 2000. Structural similarities and evolutionary relationships in chloride-dependent α-amylases. *Gene* 253:95-105.

Darling, K.F., C.M. Wade, I.A. Stewart, D. Kroon, R. Dingle, and A.J. Brown. 2000. Molecular evidence for genetic mixing of Arctic and Antarctic subpolar populations of planktonic foraminifera. *Nature* 405:43-47.

Davail, S., G. Feller, E. Narinx, and C. Gerday. 1994. Cold adaptation of proteins: Purification, characterization, and sequence of the heat-labile subtilisin from the Antarctic psychrophile *Bacillus* TA41. *Journal of Biological Chemistry* 269:17448-17453.

Davidson, A.T., H.J. Marchant, and W.K. de la Mare. 1996. Natural UVB exposure changes the species composition of Antarctic phytoplankton in mixed cultures. *Aquatic Microbial Ecology* 10:299-305.

Day, T.A., and P.J. Neale. 2002. Effects of UV-B radiation on terrestrial and aquatic primary producers. *Annual Review of Eccological Systems* 33:371-396.

de Baar, H.J.W., J.T.M. de Jong, D.C.E. Bakker, B.M. Löscher, C. Veth, U. Bathmann, and V. Smetacek. 1995. Importance of iron for phytoplankton spring blooms and $CO_2$ drawdown in the Southern Ocean. *Nature* 373:412-415.

DeConto, R.M., and D. Pollard. 2003. Rapid Cenozoic glaciation of Antarctica triggered by declining atmospheric $CO_2$. *Nature* 421:245-249.

Delong, E.F., K.Y. Wu, B.B. Prezelin, and R.V.M. Jovine. 1994. High abundance of Archaea in Antarctic marine picoplankton. *Nature* 371:695-697.

Delong, E.F., D.G. Franks, and A.A. Yayanos. 1997. Evolutionary relationships of cultivated psychrophilic and barophilic deep-sea bacteria. *Applied and Environmental Microbiology* 63:2105-2108.

Deloukas, P., G. D. Schuler, G. Gyapay, E. M. Beasley, C. Soderlund, P. Rodriguez-Tomé, L. Hui, T.C. Matise, K.B. McKusick, J.S. Beckmann, S. Bentolila, M.-T. Bihoreau, B.B. Birren, J. Browne, A. Butler, A.B. Castle, N. Chiannilkulchai, C. Clee, P., J.R. Day, A. Dehejia, T. Dibling, N. Drouot, S. Duprat, C. Fizames, et al. 1998. A physical map of 30,000 human genes. *Science* 282:744-746.

Deming, J.W. 2002. Psychrophiles and polar regions. *Current opinion. Microbiology* 3(5):301-309.

de Mora, S., S. Demers, and M. Vernet (eds.). 2000. *The Effects of UV Radiation in the Marine Environment.* Cambridge University Press, UK.

Denton, G.H., and B.L. Hall. 2000. Glacial and paleoclimatic history of the Ross ice drainage system of Antarctica: Preface. *Geografiska Annaler* 82A:139-141.

DeWitt, H.H. 1971. Coastal and deep-water benthic fishes of the Antarctic. Pp. 1-10 in V.C. Bushnell (ed.), *Antarctic Map Folio Series, Folio 15*. New York: American Geographical Society.

Diamond, J. 2001. Damned experiments! *Science* 294:1847-1848.

DiTullio, G.R., J.M. Grebmeier, K.R. Arrigo, M.P. Lizotte, D.H. Robinson, A. Leventer, J.P. Barry, M.L. Vanwoert, and R.B. Dunbar. 2000. Rapid and early export of *Phaeocystis antarctica* blooms in the Ross Sea, Antarctica. *Nature* 404:595-598.

Doran, P.T. J.C. Priscu, W. Berry Lyons, J.E. Walsh, A.G. Fountain, D.M. McKnight, D.L. Moorhead, R.A. Virginia, D.H. Wall, G.D. Clow, C.H. Fritsen, C.P. McKay, and A.N. Parsons. 2002. Antarctic climate cooling and terrestrial ecosystem response. *Nature* 415:517-520.

Duman, J.G. 2001. Antifreeze and ice nucleator proteins in terrestrial arthropods. *Annual Review of Physiology.* 63:327-357.

Eastman, J.T. 1991. Evolution and diversification of Antarctic notothenioid fishes. *American Zoologist* 31:93-109.

Eastman, J.T. 1993. *Antarctic Fish Biology: Evolution in a Unique Environment.* Academic Press, San Diego.
Eastman, J.T. 1999. Aspects of the biology of the icefish *Dacodraco hunteri* (Notothenioidei, Channichthyidae) in the Ross Sea, Antarctica. *Polar Biology* 21:194-196.
Eastman, J.T. 2000. Antarctic notothenioid fishes as subjects for research in evolutionary biology. *Antarctic Science* 12:276-287.
Eastman, J.T., and A.R. McCune. 2000. Fishes on the Antarctic continental shelf: Evolution of a marine species flock? *Journal of Fish Biology* 57A:84-102.
Emanuel, B.S., and T.H. Shaikh. 2001. Segmental duplications: An "expanding" role in genomic instability and disease. *Nature Review Genetics* 2:791-800.
Estell, D.A. 1988. Modified enzymes and methods for making same. U.S. Patent 4760025.
Falkowski, P.G., and K. Kolber, 1993 Estimation of phytoplankton photosynthesis by active fluorescence. *International Consortium on the Exploration of the Seas Symposium* 197:92-103.
Fell, D.A. 2001. Beyond genomics. *Trends in Genetics* 17:680-683.
Feller, G., and C. Gerday. 1997. Psychrophilic enzymes: Molecular basis of cold adaptation. *Cellular and Molecular Life Sciences* 53:830-841.
Feller, G., M. Thiry, J.L. Arpigny, and C. Gerday. 1991. Cloning and expression in *Escherichia coli* of three lipase-encoding genes from the psychrotrophic Antarctic strain *Moraxella* TA144. *Gene* 102:111-115.
Felsenstein, J. 1989. PHYLIP: Phylogeny inference package, Version 3.2. *Cladistics* 5:164-166.
Ferrari, V.C., and J.T. Hollibaugh. 1999. Distribution of microbial assemblages in the central Arctic Ocean basin studied by PCR/DGGE: Analysis of a large data set. *Hydrobiologia* 401:55-68.
Fiehn, O. 2001. Combining genomics, metabolome analysis, and biochemical modeling to understand metabolic networks. *Comparative and Functional Genomics* 2:155-168.
Fields, P.A., and G.N. Somero. 1998. Hot spots in cold adaptation: Localized increases in conformational flexibility in lactate dehydrogenase $A_4$ orthologs of Antarctic notothenioid fishes. *Proceedings of the National Academy of Sciences USA* 95:11,476-11,481.
Fletcher, G.L., C.L. Hew, and P.L. Davies. 2001. Antifreeze proteins of teleost fishes. *Annual Review of Physiology* 63:488-493.
Francis, J.E. 1999. Evidence from fossil plants for Antarctic paleoclimates over the past 100 million years. *Terra Antartica Reports* 3:43-52.
Frati, F., and E. Dell'Ampio. 2000. Molecular phylogeny of three subfamilies of the Neanuridae (Insecta, Collembola) and the position of the Antarctic species Friesea grisea Schaffer. *Pedobiologia* 44:342-360.
Frati, F., P.P. Fanciulli, A. Carapelli, E. Dell'Ampio, F. Nardi, G. Spinsanti, and R. Dallai. 2000. DNA sequence analysis to study the evolution of Antarctic Collembola. *Italian Journal of Zoology* 67:133-139.
Frati, F., G. Spinsanti, and R. Dallai. 2001. Genetic variation of mtCOII gene sequences in the collembolan Isotoma klovstadi from Victoria Land, Antarctica: Evidence for population differentiation. *Polar Biology* V24:934-940.
Freckman, D.W., and R.A. Virginia. 1998. Soil biodiversity and community structure in the McMurdo Dry Valleys, Antarctica. Pages 323-335 in J. Priscu (ed.), *Ecosystem Dynamics in a Polar Desert: The McMurdo Dry Valleys, Antarctica.* Washington, DC: American Geophysical Union.
Frederick, J. E. and H. E. Snell. 1988. Ultraviolet radiation levels during the Antarctic spring. *Science* 241:438-440.
Frederick, J.E., Q. Zheng, and C.R. Booth. 1998. Ultraviolet radiation at sites on the Antarctic coasts. *Journal of Photochemistry and Photobiology* 38:183-190.

Fritsen, C.H., and J.C. Priscu. 1998. Cyanobacterial assemblages in permanent ice covers of Antarctic lakes: Distribution, growth rate, and temperature response of photosynthesis. *Journal of Phycology* 34:587-597.

Funk, D.E., E.R. Pullman, K.M. Peterson, P. Crill, and W.D. Billings. 1994. The influence of water table on carbon dioxide, carbon monoxide and methane fluxes from taiga bog microcosms. *Global Biogeochemical Cycles* 8:271-278.

Gaidos, E.J., and F. Nimmo. 2000. Tectonics and water on Europa. *Nature* 405:637.

Gaidos, E.J., K.H. Nealson, and J.L. Kirschvink. 1999. Life in ice-covered oceans. *Science* 284:1631-1633.

Gasch, A.P., P.T. Spellman, C.M. Kao, O. Carmel-Harel, M.B. Eisen, G. Stgorz, D. Botstein, and P.O. Brown. 2000. Genomic expression programs in the response of yeast cells to environmental changes. *Molecular Biology of the Cell* 11(12):4241-4257.

Gast, R.J., D.A. Caron, D.M. Moran, J.M. Rose, M.R. Dennett, R.A. Schaffner, and D.J. Patterson. 2002. Protistan molecular ecology and physiology in Antarctic waters. P. 34 in Moran, M.A., and S.L. Cadee (eds.), *Microbial Observatories/LexEn Principal Investigator's Workshop*. Arlington, Virginia: National Science Foundation.

GenomeWeb Staff. 2002. NHGRI gives GenomeVision $1.6M to make gene sequencing cheaper. *GenomeWeb*, 8 October 2002. Available at <http://www.genomeweb.com>. Accessed November 4, 2003.

Georlette, D., Z.O. Jonsson, F. Van Petegem, J. Chessa, J. Van Beeumen, U. Hubscher, and C. Gerday. 2000. A DNA ligase from the psychrophile *Pseudoalteromonas haloplanktis* gives insights into the adaptation of proteins to low temperatures. *European Journal of Biochemistry* 267:3502-3512.

Gerday, C., M. Aittaleb, J.L. Arpigny, E. Baise, J.P. Chessa, G. Garsoux, I. Petrescu, and G. Feller. 1997. Psychrophilic enzymes: A thermodynamic challenge. *Biochimica et Biophysica Acta* 1342:119-131.

Gerday, C., M. Aittaleb, M. Bentahir, J.P. Chessa, P. Claverie, T. Collins, S. D'Amico, J. Dumont, G. Garsoux, D. Georlette, A. Hoyoux, T. Lonhienne, M.A. Meuwis, and G. Feller. 2000. Cold-adapted enzymes: From fundamentals to biotechnology. *Trends in Biotechnology* 18:103-107.

Gibson, G., and S.V. Muse. 2002. *A Primer of Genome Science*. Sunderland, Massachusetts: Sinauer Associates. 347 pp.

Gilbert, N., S. Lutz-Prigge, and J.V. Moran. 2002. Genomic deletions created upon LINE-1 retrotransposition. *Cell* 110:315-325.

Gilichinsky, D., E. Vorobyova, L. Erokhina, and D. Fyodorov-Davydov. 1992. Long-term preservation of microbial ecosystems in permafrost. *Advances in Space Research* 12:255-263.

Gilmour, S.J., A.M. Sebolt, M.P. Salazar, J.D. Everard, and M.F. Thomashow. 2000. Overexpression of the Arabidopsis *CBF3* transcriptional activator mimics multiple biochemical changes associated with cold acclimation. *Plant Physiology* 124:1854-1865.

Gleitz, M., M.R. Vonderloeff, D.N. Thomas, G.S. Dieckmann, and F.J. Millero. 1995. Comparison of summer and winter inorganic carbon, oxygen and nutrient concentrations in Antarctic see-ice brine. *Marine Chemistry* 51:81-91.

Gon, O., and P.C. Heemstra (eds.). 1990. *Fishes of the Southern Ocean*. J.L.B. Smith Institute of Ichthyology, Grahamstown, South Africa.

Gosink, J.J., R.L. Irgens, and J.T. Staley. 1993. Vertical distribution of bacteria in sea ice. *FEMS Microbiology Ecology* 102:85-90.

Gosink, J.J., R.P. Herwig, and J.T. Staley. 1997. *Octadecabacter arcticus* gen. nov., sp. nov., and *O. antarcticus*, sp. nov., non-pigmented, psychrophilic gas vacuolate bacteria from polar sea ice and water. *Systematic and Applied Microbiology* 20:356-365.

# REFERENCES

Gracey, A.Y., J.V. Troll, and G.N. Somero. 2001. Hypoxia-induced gene expression profiling in the euryoxic fish *Gillichthys mirabilis*. *Proceedings of the National Academy of Sciences USA* 98:1993-1998.

Gray, N.D., R. Howarth, R.W. Pickup, J.G. Jones, and I.M. Head. 2000. Use of combined microautoradiography and fluorescence in situ hybridization to determine carbon metabolism in mixed natural communities of uncultured bacteria from the genus *Achromatium*. *Applied and Environmental Microbiology* 66:4518-4522.

Greenberg, R., P. Geissler, B.R. Tufts, and G.V. Hoppa. 2000. Habitability of Europa's crust: The role of tidal-tectonic processes. *Journal of Geophysical Research* 105:17,551-17,562.

Grellmann, D. 2002. Plant responses to fertilization and exclusion of grazers on an arctic tundra heath. *OIKOS* 98:190-204.

GRIP (Greenland Ice-core Project) Members. 1993. Climate instability during the last interglacial period recorded in the GRIP ice core. *Nature* 364:203-207.

Grootes, P.M., M. Stuiver, J.W.C. White, S. Johnsen, and J. Jouzel. 1993. Comparison of oxygen isotope records from the GISP2 and GRIP Greenland ice cores. *Nature* 366:552-554.

Guschin, D.Y., B.K. Mobarry, D. Proudnikov, D.A. Stahl, B.E. Rittmann, and A.M. Mirzabekov. 1997. Oligonucleotide microchips as genosensors for determinative and environmental studies in microbiology. *Applied and Environmental Microbiology* 63:2397-2402.

Haake, V., D. Cook, J.L. Riechmann, O. Pineda, M.F. Thomashow, and J.Z. Zhang. 2002. Transcription factor CBF4 is a regulator of drought adaptation in Arabidopsis. *Plant Physiology* 130:639-948.

Have, P., J. Nielsen, and A. Botner. 1991. The seal death in Danish waters 1988. 2. Virological studies. *Acta Veterinaria Scandinavica* 32:211-219.

Herbert, R.A. 1992. A perspective on the biotechnological potential of extremophiles. *Trends in Biotechnology* 10:395-402.

Hewitt, G.M. 2000. The genetic legacy of the Quaternary ice ages. *Nature* 405:907-913.

Higuchi, R., B. Bowman, M. Freiberger, O. Ryder, and A. Wilson. 1984. DNA sequence from the quagga, an extinct member of the horse family. *Nature* 312:282-84.

Hobbie, J.E. 1996. Temperature and plant species control over litter decomposition in Alaskan tundra. *Ecological Monograph* 66:503-522.

Hobbie, J.E. (ed.). 1980. *Limnology of Tundra Ponds: Barrow, Alaska*. Stroudsburg, Penn.: Dowden Hutchinson, & Ross. 514 pp.

Hochachka, P.W. 1988. Channels and pumps—determinants of metabolic cold adaptation strategies. *Comparative Biochemistry and Physiology* 90B:515-519.

Hochachka, P.W., and G.N. Somero. 2002. *Biochemical Adaptation: Mechanism and Process in Physiological Evolution*. Oxford University Press, New York. 466 pp.

Hofmann, D.J. 1996. The 1996 Antarctic ozone hole. *Nature* 383:129.

Hofmann, G.E., and G.N. Somero. 1995. Evidence for protein damage at environmental temperatures: Seasonal changes in levels of ubiquitin conjugates and hsp70 in the intertidal mussel *Mytilus trossulus*. *Journal of Experimental Biology* 198:1509-1518.

Hofmann, G.E., B.A. Buckley, S. Airaksinen, J.E. Keen, and G.N. Somero. 2000. Heat-shock protein expression is absent in the Antarctic fish *Trematomus bernacchii* (family Nototheniidae). *Journal of Experimental Biology* 203:2331-2339.

Hofreiter, M., D. Serre, H.N. Poinar, M. Kuck, and S. Paabo. 2001. Ancient DNA. *Nature Reviews/Genetics* 2:353-359.

Hollibaugh, J.T., N. Bano, and H.W. Ducklow. 2002. Widespread distribution in polar oceans of a 16S rRNA gene sequence with affinity to *Nitrosospira*-like ammonia-oxidizing bacteria. *Applied and Environmental Microbiology* 68:1478-1484.

Hoyoux, A., I. Jennes, P. Dubois, S. Genicot, F. Dubail, J. M. Francois, E. Baise, G. Feller, and C. Gerday. 2001. Cold-adapted β-galactosidase from the Antarctic psychrophile *Pseudoalteromonas haloplanktis*. *Applied and Environmental Microbiology* 67:1529-1535.

Hugenholtz, P., B.M. Goebel, and N.R. Pace. 1998. Impact of culture-independent studies on the emerging phylogenetic view of bacterial diversity. *Journal of Bacteriology* 180(18):4765-4774.

Huner, P.A., G. Öquist, and F. Sarhan. 1998. Energy balance and acclimation to light and cold. *Trends in Plant Science* 3(6):224-230.

Huston, A., and J. Deming. 2002. Relationships between microbial extracellular enzymatic activity and suspended and sinking particulate organic matter: Seasonal transformations in the North Water. *Deep-Sea Research Part II-Topical Studies in Oceanography* 49:5211-5225.

Imai, R., L. Chang, A. Ohta, E.A. Bray, and M. Takagi. 1996. A LEA-class gene of tomato confers salt and freezing tolerance when expressed in *Saccharomyces cerevisiae*. *Gene* 170:243-248.

Irgens, R.L., I. Suzuki, and J.T. Staley. 1989. Gas vacuolate bacteria obtained from marine waters of Antarctica. *Current Microbiology* 18:261-265.

Isachenko, B. 1912. Some data on permafrost bacteria. *Izvestiya Sankt Peterburgskogo Botanicheskogo Sada* 12:5-6.

Ivanov, A.G., E. Miskiewicz, A.K. Clarke, B.M. Greenburg, and N.P.A. Hüner. 2000. Protection of Photosystem II against UV-A and UV-B radiation in the cyanobacterium *Plectonema boryanum*: The role of growth temperature and growth irradiance. *Journal of Photochemistry and Photobiology* 72(6):772.

Jaglo-Ottosen, K.R., S.J. Gilmour, D.G. Zarka, O. Schabenberger, and M.F. Thomashow. 1998. *Arabidopsis CBF1* overexpression induces *COR* genes and enhances freezing tolerance. *Science* 280:104-106.

Jain, R., M.C. Rivera, J.E. Moore, and J.A. Lake. 2002. Horizontal gene transfer in microbial genome evolution. *Theoretical Population Biology* 61:489-495.

Jeffrey, W.H., P. Aas, M. Maille Lyons, R.B. Coffin, R.J. Pledger, and D.L. Mitchell. 1996. Ambient solar-radiation induced photodamage in marine bacterioplankton. *Journal of Photochemistry and Photobiology* 64:419-427.

Jeffrey, W.H., J.P. Kase, and S.W. Wilhelm. 2000. Ultraviolet radiation effects on bacterioplankton and viruses in marine ecosystems. Pp 206-236 in S.J. De Mora et al. (eds.), *Effects Of UV Radiation On Marine Ecosystems*. United Kingdom: Cambridge University Press.

Jiang, W., Y. Hou, and M. Inouye. 1997. CspA, the major cold-shock protein of *Escherichia coli*, is an RNA chaperone. *Journal of Biological Chemistry* 272:196-202.

Johannessen, O.L., E.V. Shalina, and M.W. Miles. 1999. Satellite evidence for an Arctic sea ice cover in transformation. *Science* 286:1937-1939.

Johnsen, S. J., H. B. Clausen, W. Dansgaard, K. Fuhrer, N. Gundestrup, C. U. Hammer, P. Iversen, J. Jouzel, B. Stauffer, and J. F. Steffensen. 1992. Irregular glacial interstadials recorded in a new Greenland ice core. *Nature* 359:311-313.

Jones, A.E., and J.D. Shanklin. 1995. Continued decline of total ozone over Halley, Antarctica, since 1985. *Nature* 376:409-411.

Kaeberlin, T., K. Lewis, and S.S. Epstein. 2002. Isolating "uncultivable" microorganisms in pure culture in a simulated natural environment. *Science* 296:1127-1129.

Karentz, D., and I. Bosch. 2001. Influence of ozone-related increases in ultraviolet radiation on antarctic marine organisms. *American Zoologist* 41:3-16.

Karentz, D., J.E. Cleaver, and D.L. Mitchell. 1991. Cell survival characteristics and molecular responses of Antarctic phytoplankton to ultraviolet-B radiation. *Journal of Phycology* 27:326-341.

# REFERENCES

Karl, D.M., D.F. Bird, K. Bjorkman, T. Houlihan, R. Shackelford, and L. Tupas. 1999. Microorganisms in the accreted ice of Lake Vostok, Antarctica. *Science* 286:2144-2147.

Kasuga, M., Q. Liu, S. Miura, K. Yamaguchi-Shinozaki, and K. Shinozaki. 1999. Improving plant drought, salt, and freezing tolerance by gene transfer of a single stress-inducible transcription factor. *Nature Biotechnology* 17:287-291.

Kawall, H.G., J.J. Torres, B.D. Sidell, and G.N. Somero. 2002. Metabolic cold adaptation in Antarctic fishes: Evidence from enzymatic activities in brain. *Marine Biology* 140:279-286.

Kazazian, H.H., Jr., and J.L. Goodier. 2002. LINE drive: Retrotransposition and genome instability. *Cell* 110:277-280.

Kennett, J.P. 1977. Cenozoic Evolution of Antarctic glaciation, the circum-Atlantic ocean, and their impact on global paleoceanography. *Journal of GeophysicalResearch* 82:3843-3859.

Kennett, J.P. 1982. Marine Geology. Chapter 19. *Global paleoceanographic evolution: Critical events in ocean history.* Englewood Cliffs, NJ: Prentice Hall. Pp. 696-751.

Kerr, R.A. 1999. Will the Arctic Ocean lose all its ice? *Science* 286:1828.

Kerry, K.R., and G. Hempel (eds.). 2002. *Antarctic Ecosystems: Ecological Change and Conservation.* Springer-Verlag, Berlin and Heidelberg.

Kivelson, M.G., K.K. Khurana, C.T. Russell, M. Volwerk, R.J. Walker, and C. Zimmer. 2000. Galileo magnetometer measurements: A stronger case for a subsurface ocean at Europa. *Science* 289:1340-1343.

Kolber Z.S., R.T. Barber, K.H. Coale, S.E. Fitzwater, R.M. Greene, K.S. Johnson, S. Lindley, and P.G. Falkowski. 1994. Iron limitation of phytoplankton photosynthesis in the equatorial Pacific Ocean. *Nature* 371:145-149.

Komarkova, V., S. Poncet, and J. Poncet. 1985. Two native Antarctic vascular plants, Deschampsia antarctica and Colobanthus quitensis in the Antarctic Peninsula area. *Arctic and Alpine Research* 122:108-113.

Koonin, E.V., K.S. Makarova, and L. Aravind. 2001. Horizontal gene transfer in prokaryotes: Quantification and classification. *Annual Review of Microbiology* 55:709-742.

Krupnik, I., and D. Jolly, (eds.). 2002. *The Earth is faster now: Indigenous observations of Arctic environmental change.* Fairbanks, AK: Arctic Research Consortium of the United States. 384 pp.

Kwok, C., R. Critcher, and K. Schmitt. 1999. Construction and characterization of zebrafish whole genome radiation hybrids. Pp. 287-302 in H.W. Detrich III, M. Westerfield, and L.I. Zon (eds.), *Methods in Cell Biology, Vol. 60, "The Zebrafish: Genetics and Genomics."* San Diego: Academic Press.

Lambert, D.M., P.A. Richie, C.D. Millar, B. Holland, A.J. Drummond, and C. Baroni. 2002. Rates of evolution in ancient DNA from Adélie Penguins. *Science* 295:2270-2273.

Lander, E.S., L.M. Linton, B. Birren, C. Nusbaum, M.C. Zody, J. Baldwin, K. Devon, K. Dewar, M. Doyle, W. Fitzhugh, R. Funke, D. Gage, K. Harris, A. Heaford, J. Howland, L. Kann, J. Lehoczky, R. Levine, P. Mcewan, K. Mckernan, J. Meldrim, J.P. Mesirov, C. Miranda, W. Morris, J. Naylor, C. Raymond, M. Rosetti, R. Santos, A. Sheridan, C. Sougnez, N. Stange-Thomann, N. Stojanovic, A. Subramanian, and D. Wyman, et al. 2001. Intitial sequencing and analysis of the human genome. *Nature* 409:860-921.

Lakhman. K. 2002. New Venter center will likely be a boon for overall genome sequencing. *GenomeWeb*, 16 August 2002. Available at <http://www.genomeweb.com>. Accessed November 3, 2002.

Langseth, M., T. Delaca, G. Newton, B. Coakley, R. Colony, J. Gossett, et al. 1993. SCICEX 93 Arctic Cruise of the United States Navy, nuclear-powered submarine, USS Pargo. *Marine Technology Society Journal* 27:4-12.

Lau, D.T., A. Saeed-Kothe, S.K. Parker, and H.W. Detrich III. 2001. Adaptive evolution of gene expression in Antarctic fishes: Divergent transcription of the 5'-to-5' linked adult α1- and β-globin genes of the Antarctic teleost *Notothenia coriiceps* is controlled by dual promoters and intergenic enhancers. *American Zoologist* 41:113-132.

Lawver, L.A., L.M. Gahagan, and M.F. Coffin. 1992. The development of paleoseaways around Antarctica. Pp. 7-30 in Kennett, J.P. and D.A. Warnke (eds.), *The Antarctic Paleoenvironment: A Perspective on Global Change.* Antarctic Research Series Volume 56. Washington, DC: American Geophysical Union.

Lee, N, P.H. Nielsen, K.H. Andreasen, S. Juretschko, J.L. Nielsen, K.H. Schleifer, and M. Wagner. 1999. Combination of fluorescent *in situ* hybridization and microautoradiography—a new tool for structure-function analyses in microbial ecology. *Applied and Environmental Microbiology* 65:1289-1297.

Leonard, J.A., R.K. Wayne, and A. Cooper. 2000. Population genetics of Ice Age brown bears. *Proceedings of the National Academy of Sciences USA* 97:1651-1654.

Lesser, M.P., J.J. Cullen, and P.J. Neale. 1994. Carbon uptake in a marine diatom during acute exposure to ultraviolet B radiation: Relative importance of damage and repair. *Journal of Phycology* 30:183-192.

Levi, B.G. 1998. Adrift on the ice pack, researchers explore changes in the Arctic environment. *Physics Today* 51:17-19.

Lewis-Smith, R.I. 1996. Introduced plants in Antarctica: Potential impacts and conservation issues. *Biological Conservation* 76:135-146.

Lindahl, T. 1993. Instability and decay of the primary structure of DNA. *Nature* 362:709-715.

Litchman, E., P.J. Neale, and A.T. Banaszak. 2002. Increased sensitivity to ultraviolet radiation in nitrogen-limited dinoflagellates: Photoprotection and repair. *Limnology and Oceanography* 47:86-94.

Littlepage, J.L. 1965. Oceanographic investigation in McMurdo Sound, Antarctica. Pp. 1-37 in Llano, G.A. (ed.), *Biology of the Antarctic Seas.* Volume II. Washington, DC: American Geophysical Union.

Liu, Q., M. Kasuga, Y. Sakuma, H. Abe, S. Miura, K. Yamaguchi-Shinozaki, and K. Shinozaki. 1998. Two transcription factors, DREB1 and DREB2, with an EREBP/AP2 DNA binding domain separate two cellular signal transduction pathways in drought- and low-temperature-responsive gene expression, respectively, in Arabidopsis. *Plant Cell* 10:1391-1406.

Lizotte, M.P., and J.C. Priscu. 1992. Spectral irradiance and bio-optical properties in perennially ice-covered lakes of the dry valleys (McMurdo Sound, Antarctica). *Antarctic Research Series* 57:1-14.

Logsdon, J.M., Jr., and W.F. Doolittle. 1997. Origin of antifreeze protein genes: A cool tale in molecular evolution. *Proceedings of the National Academy of Sciences USA* 94:3485-3487.

Lonhienne, T., J. Zoidakis, C. E. Vorgias, G. Feller, C. Gerday, and V. Bouriotis. 2001. Modular structure, local flexibility and colc-activity of a novel chitobiase from a psychrophilic Antarctic bacterium. *Journal of Molecular Biology* 310:291-297.

Lubin, D., J.E. Frederick, C.R. Booth, T. Lucas, and D. Neuschuler. 1989. The ultraviolet radiation environment of Antarctica: McMurdo Station during September–October 1987. *Geophysical Research Letters* 16:783-785.

Malin, M.C., and M.H. Carr. 1999. Groundwater formation of martian valleys. *Nature* 397:589-591.

Malloy, K.D., M.A. Holman, D. Mitchell, and H.W. Detrich III. 1997. Solar UVB-induced DNA damage and photoenzymatic repair in antarctic zooplankton. *Proceedings of the National Academy of Sciences USA* 94:1258-1263.

# REFERENCES

Marchant, H.J., A.T. Davidson, and G.J. Kelly. 1991. UV-B protecting compounds in the marine alga *Phaeocystis pouchetii* from Antarctica. *Marine Biology* 109:391-395.

Marchant, D.R., C.C. Swisher III, D.R. Lux, D.P. West, Jr., and G.H. Denton. 1993. Pliocene paleoclimate and East Antarctic ice-sheet history from durficial ash deposits. *Science* 260:667-670.

Marsh, A.G., R.E. Maxon, Jr., and D.T. Manahan 2001. High macromolecular synthesis with low metabolic cost in Antarctic sea urchin embryos. *Science* 291:1950-1952.

Martens, K. 1997. Speciation in ancient lakes. *Trends in Ecology and Evolution* 12:177-182.

Martinez, J.G., and F.R. Valera. 2000. Microdiversity of uncultured marine prokaryotes: The SAR11 cluster and the marine Archaea of Group I. *Molecular Ecology* 9:935-948.

Masri, S., H. Rast, T. Ripley, D. James, M. Green, X Jia, and R.H. Devlin. 2002. Detection of genetically modified coho salmon using polylerase chain reaction (PCR) amplification. *Journal of Agricultural and Food Chemistry* 50:3161-3164.

Massana, R., L.T. Taylor, A.E. Murray, K.Y. Wu, W.H. Jeffrey, and E.F. DeLong. 1998. Vertical distribution and temporal variation of marine planktonic archaea in the Gerlache Strait, Antarctica, during early spring. *Limnology and Oceanography* 43:607-617.

Maxwell, D.P., S. Falk, C.G. Trick, and N.P.A. Huner. 1994. Growth at low temperature mimics high-light acclimation in *Chlorella vulgaris*. *Plant Physiology* 105:535-543.

Maxwell, D.P., S. Falk, and N.P.A. Huner. 1995. Photosystem II excitation pressure and development of resistance to photoinhibition. I. Light harvesting complex II abundance and zeaxanthin content in Chlorella vulgaris. *Plant Physiology* 107:687-694.

McCune, A.R. 1997. How fast is speciation? Molecular, geological, and phylogenetic evidence from adaptive radiation of fishes. Pp. 585-610 in T.J. Givnish and K.J. Sytsma (eds.), *Molecular Evolution and Adaptive Radiation*. United Kingdom: Cambridge University Press.

McGowan, K. 2002a. Alternative sequencing technologies. *GenomeWeb*, 1 October 2002. Available at <http://www.genomeweb.com>. Accessed November 3, 2002.

McGowan, K. 2002b. For TIGR and Venter, the goal now is to sequence entire ecosystems at a time. *GenomeWeb*, 22 July 2002. Available at www.genomeweb.com. Accessed October 30, 2002.

McKay, C.P. 2001. The deep biosphere: Lessons for planetary exploration. Pp. 315-327 in J.K. Fredrickson and M. Fletcher (eds.), *Subsurface Microbiology and Biogeochemistry*. New York: Wiley-Liss, Inc.

McKay, C.P., and C.R. Stoker. 1989. The early environment and its evolution on Mars: Implications for life. *Reviews of Geophysics* 27:189-214.

McKay, D.S., E.K. Gibson, K.L. Thomas-Keptra, H. Vali, S. Romanek, S.J. Clemett, X.D.F. Chillier, C.R. Maechling, and N. Zare. 1996. Search for past life on Mars: Possible relic biogenic activity in Martian meteorite ALH84001. *Science* 273:924-930.

Meador, J., W.H. Jeffrey, J.P. Kase, J.D. Pakulski, S. Chiarello, and D.L. Mitchell. 2002. Seasonal fluctuation of DNA photodamage in marine plankton assemblages at Palmer Station, Antarctica. *Journal of Photochemistry and Photobiology* 75:266-271.

Medlin, L.K., M. Lange, and M.E.M. Baumann. 1994. Genetic differentiation among three colony-forming species of *Phaeocystis*: Further evidence for the phylogeny of the Prymnesiophyta. *Phycologia* 33:199-212.

Michaelis, W., R. Seifert, K. Nauhaus, T. Treude, V. Thiel, M. Blumenberg, K. Knittel, A. Gieseke, K. Peterknecht, T. Pape, A. Boetius, R. Amann, B.B. Jørgensen, F. Widdel, J. Peckmann, N.V. Pimenov, and M.B. Gulin. 2002. Microbial reefs in the Black Sea fueled by anaerobic oxidation of methane. *Science* 297:1013-1015.

Miller, D.N., J.E. Bryant, E.L. Madsen, and W.C. Ghiorse. 1999. Evaluation and optimization of DNA extraction and purification procedures for soil and sediment samples. *Applied and Environmental Microbiology* 65:4715-4724.

Moisan, T.A., and B.G. Mitchell. 2001. UV absorption by mycosporine-like amino acids in *Phaeocystis antarctica* Karsten induced by photosynthetically available radiation. *Marine Biology* 138:217-227.

Molina, L.T., and M.J. Molina. 1986. Absolute absorption cross-sections of ozone in the 185- to 350-nm wavelength range. *Journal of Geophysical Research and Atmospheres* 91:14501-14508.

Morgan, R.M., A.G. Ivanov, J.C. Priscu, D. P. Maxwell, and N.P.A. Huner. 1998. Structure and composition of the photochemical apparatus of the Antarctic green alga, *Chlamydomonas subcaudata*. *Photosynthesis Research* 56:303-314.

Moritz, R.E., C.M. Blitz, and E.J. Steig. 2002. Dynamics of recent climate changes in the Arctic. *Science* 207:1497-1502.

Mostajir, B., S. Demers, S. de Mora, R.P. Bukata, and J.H. Jerome. 2000. Implications of UV radiation for the food web structure and consequences on the carbon flow. Pp. 311-320 in S. de Mora, S. Demers and M. Vernet (eds.), *The Effects of UV Radiation in the Marine Environment*. United Kingdom: Cambridge University Press.

Moylan, T.J., and B.D. Sidell. 2000. Concentrations of myoglobin and myoglobin mRNA in heart ventricles from Antarctic fishes. *Journal of Experimental Biology* 203:1277-1286.

Mueller, U.G., and L.L. Wolfenbarger. 1999. ALFP genotyping and fingerprinting. *Trends in Ecology and Evolution* 14:389-394.

Mullis, K.B., and F.A. Faloona. 1987. Specific synthesis of DNA in vitro via a polymerase catalyzed chain reaction. *Methods in Enzymology* 155:335-350.

Murray, A.E., C.M. Preston, R. Massana, L.T. Taylor, A. Blakis, K.Y. Wu, and E.F. DeLong. 1998. Seasonal and spatial variability of bacterial and archaeal assemblages in the coastal waters near Anvers Island, Antarctica. *Applied and Environmental Microbiology* 64:2585-2595.

Murray, A.E., K.Y. Wu, C.L. Moyer, D.M. Karl, and E.F. DeLong. 1999. Evidence for circumpolar distribution of planktonic Archaea in the Southern Ocean. *Aquatic Microbial Ecology* 18:263-273.

Murry, R.E., K. E. Cooksey, and J.C. Priscu. 1986. Stimulation of bacterial DNA synthesis by algal exudates in an attached algal-bacterial consortium. *Applied and Environmental Microbiology* 52:1177-1182.

Nadeau, T.-L., and R.W. Castenholz. 2000. Characterization of psychrophilic oscillatorians (cyanobacteria) from Antarctic meltwater ponds. *Journal of Phycology* 36:914-923.

Nadeau, T.L., E.C. Milbrandt, and R.W. Castenholz. 2001. Evolutionary relationships of cultivated Antarctic oscillatorians (Cyanobacteria). *Journal of Phycology* 37:650-654.

Narinx, E., E. Baise, and C. Gerday. 1997. Subtilisin from psychrophilic Antarctic bacteria: Characterization and site-directed mutagenesis of residues possibly involved in the adaptation to cold. *Protein Engineering* 10:1271-1279.

Nasevicius, A., and C.S. Ekker. 2000. Effective targeted gene "knockdown" in zebrafish. *Nature Genetics* 26:216-220.

NRC (National Research Council). 1991. *Opportunities and Priorities in Arctic Geoscience*. Washington, DC: National Academy Press.

NRC. 1999. *Our Common Journey: A Transition Towards Sustainability*. National Academy Press, Washington DC

NRC. 2001. *Grand Challenges in Environmental Sciences*. National Academy Press, Washington, DC.

NRC. 2002. *Abrupt Climate Change: Inevitable Surprises*. Washington, DC: National Academy Press.

NSB (National Science Board). 2000. *Environmental Science and Engineering for the 21st Century: The Role of the National Science Foundation*. National Science Foundation, Arlington, Virginia.

# REFERENCES

NSF (National Science Foundation). 2000. Ecological Genomics: *The Application of Genomic Sciences to Understanding the Structure and Function of Marine Ecosystems*. Report of a Workshop on Marine Microbial Genomics to Develop Recommendations for the National Science Foundation, held April 19-20, 2000. Arlington, Virginia: National Science Foundation.

NSF ACERE (National Science Foundation Advisory Committee for Environmental Research and Education). 2003. *Complex Environmental Systems: Synthesis for Earth, Life, and Society in the 21st Century.* A 10-Year Outlook for the National Science Foundation. National Science Foundation, Washington, DC.

Neale, P.J., and J.C. Priscu. 1995. The photosynthetic apparatus of phytoplankton from an ice-covered Antarctic lake: Acclimation to an extreme shade environment. *Plant and Cell Physiology* 36:253-263.

Neale, P.J., and J.C. Priscu. 1998. Fluorescence quenching in phytoplankton of the McMurdo dry valley lakes (Antarctica): Implications for the structure and function of the photosynthetic apparatus. Vol. 72, pp. 241-254 in J.C. Priscu (ed.), *Ecosystem Dynamics in a Polar Desert, The McMurdo Dry Valley, Antarctica*. Washington, DC: American Geophysical Union.

Neale, P.J., J.J. Cullen, and R.F. Davis. 1998. Inhibition of marine photosynthesis by ultraviolet radiation: Variable sensitivity of phytoplankton in the Wedell-Scotia Confluence during the austral spring. *Limnology and Oceanography* 43:443-448.

Neale, P.J., J.J. Fritz, and Richard F. Davis. 2001. Effects of UV on photosynthesis of Antarctic phytoplankton: Models and their application to coastal and pelagic assemblages. *Revista Chilena de Historia Natural* 74:283-292.

Neff, J.C., A.R. Townsend, G. Gleixner, S.J. Lehman, J. Turnbull, and W.D. Bowman. 2002. Variable effects of nitrogen addition on stability of carbon. *Nature* 419:915-917.

Nelson, D.M., D.J. Demaster, R.B. Dunbar, and W.O. Smith, Jr. 1996. Cycling of organic carbon and biogenic silica in the Southern Ocean: Estimates of water-column and sedimentary fluxes on the Ross Sea continental shelf. *Journal of Geophysical Research* 101c:18519-18532.

Nelson, K.E., R.A. Clayton, S.R. Gill, M.L. Gwinn, R.J. Dodson, D.H. Haft, et al. 1999. Evidence for lateral gene transfer between Archaea and bacteria from genome sequence of *Thermotoga maritima*. *Nature* 399:323-329.

Niggemyer, A., S. Spring, E. Stackebrandt, and R.F. Rosenzweig. 2001. Isolation and characterization of a novel As(V)-reducing bacterium: Implications for arsenic mobilization and the genus *Desulfitobacterium*. *Applied and Environmental Microbiology* 67:5568-5580.

Olson, J.B., T.F. Steppe, R.W. Litaker, and H.W. Paerl. 1998. $N_2$-fixing microbial consortia associated with the ice cover of Lake Bonney, Antarctica. *Microbial Ecology* 36:231-238.

Olson, R.J., H.M. Sosik, A.M. Chekalyuk, and A. Shalapyonok. 2000. Effects of iron enrichment on phytoplankton in the Southern Ocean during late summer: Active fluorescence and flow cytometric analyses. *Deep-Sea Research II* 47:3179-3200.

Omelyansky, V. 1911. Bacteriological investigation of Sanga mammoth and nearby soil. *Arkhiv Biologicheskikh Nauk* 16:335-340.

Oremland, R.S., S.E. Hoeft, N. Bano, R.A. Hollibaugh, and J.T. Hollibaugh. 2002. Anaerobic oxidation of arsenite in Mono Lake water and by a facultative chemoautotroph, strain MLHE-1. *Applied and Environmental Microbiology* 68:4795-4802.

Orton, G.S., J.R. Spencer, L.D. Travis, T.Z. Martin, and L.K. Tamppari. 1996. Galileo photopolarimeter-radiometer observations of Jupiter and the Galilean satellites. *Science* 274:389-391.

Ostertag, E.M., and H.H. Kazazian, Jr. 2001. Biology of mammalian L1 retrotransposons. *Annual Review of Genetics* 35:501-538.

Ouverney, C.C., and J.A. Fuhrman. 2000. Marine planktonic Archaea take up amino acids. *Applied and Environmental Microbiology* 66:4829-4833.
Pääbo, S. 1989. Ancient DNA: Extraction, characterization, molecular cloning and enzymatic amplification. *Proceedings of the National Academy of Sciences USA* 86:1939-1943.
Paerl, H.W., and J.C. Priscu. 1998. Microbial phototrophic, heterotrophic and diazotrophic activities associated with aggregates in the permanent ice cover of Lake Bonney, Antarctica. *Applied Microbial Ecology* 36:221-230.
Pakker, H., R.S. T. Martins, P. Boelen, and A.G.J. Buma. 2000. Effects of temperature on the photoreactivation of ultraviolet-B-induced DNA damage in *Palmaria palmata* (Rhodophyta). *Journal of Phycology* 36:334-341.
Pakulski, J.D., S. Durkin, J. Kase, J. Meador, K. McClery, and W.H. Jeffrey. In preparation. Responses of Antractic heterotrophic bacteria to seasonal changes in solar irradiance.
Park, W., P. Padmanabhan, S. Padmanabhan, G.J. Zylstra, and E.L. Madsen. 2002. nahR, encoding a LysR-type transcriptional regulator, is highly conserved among naphthalene-degrading bacteria isolated from a coal tar waste-contaminated site and in extracted community DNA. *Microbiology* 148:2319-2329.
Parker, S.K., and H. W. Detrich, III. 1998. Evolution, organization, and expression of α-tubulin genes in the Antarctic fish *Notothenia coriiceps*: Adaptive expansion of a gene family by recent gene duplication, inversion, and divergence. *Journal of Biological Chemistry* 273:34358-34369.
Patarnello, T., L. Bargelloni, V. Varotto, and B. Battaglia. 1996. Krill evolution and the Antarctic ocean currents: Evidence of vicariant speciation as inferred by molecular data. *Marine Biology* 126:603-608.
PCAST (President's Committee of Advisers on Science and Technology). 1998. *Teaming with Life: Investing in Science to Understand and Use America's Living Capital*. Office of Science and Technology Policy, Washington, DC.
Pennisi, E. 2002. Recharged field's rallying cry: Gene chips for all organisms. *Science* 297:1985-1986.
Perna, N.T., G. Plunkett, V. Burland, B. Mau, J.D. Glasner, D.J. Rose, G.F. Mayhew, P.S. Evans, J. Gregor, H.A. Kirkpatrick, G. Pósfai, J. Hackett, S. Klink, A. Boutin,Y. Shao, L. Miller, E.J. Grotbeck, N.W. Davis, A. Lim, E.T. Dimalanta, K.D. Potamousis, J. Apodaca, T.S. Anantharaman, J. Lin, G. Yen, D.C. Schwartz, R.A. Welch, and F.R. Blattner. 2001. Genome sequence of enterohemmorrhagic *Escherichia coli* O157:H7. *Nature* 409:529-533.
Peterson, D., and C. Howard-Williams (eds.). 2000. *The Latitudinal Gradient Project. Antarctica New Zealand*, special publication. Christchurch, New Zealand: International Antarctic Centre.
Petit, J.-R., J. Jouzel, D. Raynaud, N. I. Barkov, J. M. Barnola, I. Basile, M. Benders, J. Chappellaz, M. Davis, G. Delaygue, M. Delmotte, V. M. Dotlyakov, M. Legrand, V. Y. Lipendoc, C. Lorius, L. Pepin, C. Ritz, E. Saltzman, and M. Stievenard. 1999. Climate and atmospheric history of the past 420,000 years from the Vostok ice core, Antarctica. *Nature* 399:429-436.
Phelps, T.J., A.V. Palumbo, and A.S. Beliaev. 2002. Metabolomics and microarrays for improved understanding of phenotypic characteristics controlled by both genomics and environmental constraints. *Current Opinion in Biotechnology* 13:20-24.
Pörtner, H.-O. 2002. Climate variations and the physiological basis of temperature dependent biogeography: system to molecular hierarchy of thermal tolerance in animals. *Comparative Biochemistry and Physiology* A 132:739-761.
Prézelin, B.B., N.P. Boucher, and R.C. Smith. 1994. Marine primary production under the influence of the Antarctic ozone hole: Icecolors '90. *Antarctic Research Series* 62:159-186.

Prézelin, B.B., M.A. Moline, and H.A. Matlick. 1998. Icecolors '93: Spectral UV radiation effects on Antarctic frazil ice algae. *Antarctic Research Series* 73:45-83.
Price, B.P. 2000. A habitat for psychrophiles in deep Antarctic ice. *Proceedings of the National Academy of Sciences USA* 97:1247-1251.
Priscu, J.C., and B. Christner. In Press. Earth's icy biosphere. In A.T. Bull (ed.), *Microbial Diversity and Prospecting*. American Society of Microbiology.
Priscu, J.C., C.H. Fritsen, E.E. Adams, S.J. Giovannoni, H.W. Paerl, C.P. McKay, P.T. Doran, D.A. Gordon, B.D. Lanoil, and J.L. Pinckney. 1998. Perennial Antarctic lake ice: An oasis for life in a polar desert. *Science* 280:2095-2098.
Priscu, J.C., E.E. Adams, W.B. Lyons, M.A. Voytek, D.W. Mogk, R.L. Brown, C.P. McKay, C.D. Takacs, K.A. Welch, C.F. Wolf, J.D. Kirstein, and R. Avci. 1999a. Geomicrobiology of sub-glacial ice above Vostok Station. *Science* 286:2141-2144.
Priscu, J.C., C.F. Wolf, C.D. Takacs, C.H. Fritsen, J. Laybourn-Parry, E.C. Roberts, and W.B. Lyons. 1999b. Carbon transformations in the water column of a perennially ice-covered Antarctic Lake. *Bioscience* 49:997-1008.
Priscu, J.C., C.H. Fritsen, E.E. Adams, H.W. Paerl, J.T. Lisle, J.E. Dore, C.F. Wolf, and J. Milucki. In press. Perennial Antarctic lake ice: A refuge for cyanobacteria in an extreme environment. In S.O. Rogers and J. Castello (eds.), *Life in Ancient Ice*. New Jersey: Princeton University Press.
Purkhold, U., A. Pommerening-Roser, S. Juretschko, M.C. Schmid, H.-P. Koops, and M. Wagner. 2000. Phylogeny of all recognized species of ammonia oxidizers based on comparative 16S rRNA and *amoA* sequence analysis: Implications for molecular diversity surveys. *Applied and Environmental Microbiology* 66:5368-5382.
Radajewski S., P. Ineson, N.R. Parekh, and J.C. Murrell. 2000. Stable-isotope probing as a tool in microbial ecology. *Nature* 403:646-649.
Ranjard L., F. Poly, and S. Nazaret. 2000. Monitoring complex bacterial communities using culture-independent molecular techniques: Application to soil environment. *Research in Microbiology* 151:167-177.
Rappé, M.S., S.A. Connon, K.L. Vergin, and S.J. Giovannoni. 2002. Cultivation of the ubiquitous SAR11 marine bacterioplankton clade. *Nature* 418:630-633.
Raymo, M.E., K. Ganley, S. Carter, D.W. Oppo, and J. McManus. 1998. Millennial-scale climate instability during the early Pleistocene epoch. *Nature* 392:699-702.
Raymond, J.A. 1993. Glycerol and water balance in a near-isosmotic teleost, winter-acclimatized rainbow smelt. *Canadian Journal of Zoology* 71:1849-1854.
Renger, G., and U. Schreiber. 1986. Practical applications of fluorometric methods to algae and higher plant research. In: *Light Emission by Plants and Bacteria*. London: Academic Press.
Riddle, D., T. Blumethal., B. Meyer, and J. Preiss. 1997. *C. elegans* II. New York: Cold Springs Harbor Laboratory Press.
Riegger, L., and D. Robinson. 1997. Photoinduction of UV-absorbing compounds in Antarctic diatoms and *Phaeocystis antarctica*. *Marine Ecology Progress Series* 160:13-25.
Ritchie, P.A., L. Bargelloni, A. Meyer, J. A. Taylor, J. A. Macdonald, and D. A. Lambert. 1996. Mitochondrial phylogeny of trematomid fishes (Nototheniidae, Perciformes) and the evolution of Antarctic fish. *Molecular Phylogenetics and Evolution* 5:383-390.
Robinson, C.H. 2002. Controls on decomposition and soil nitrogen availability at high latitudes. *Plant Soil* 242:65-81.
Robinson, C.H., P.A. Wookey, A.N. Parsons, J.A. Potter, T.V. Callaghan, J.A. Lee, M.C. Press, and J.M. Welker. 1995. The response of plant litter decomposition and nitrogen mineralisation to simulated environmental change in a high Arctic polar semi dessert and subarctic dwarf shrub heath. *Oikos* 74:503-512.

Rogers, S.O., W.T. Starmer, and J.D. Castello. In press. Recycling of organisms and genomes. In J. Castello and S. Rogers (eds.), *Life in Ancient Ice*. Princeton Press.

Rogers, S.O., and M.A.M. Rogers. 1999. Gene flow in fungi. Pp. 97-121 in J.J. Worrall (ed.), *Structure and Dynamics of Fungal Populations*. Kluwer Academic Publishers, Boston.

Rondon, M.R., P.R. August, A.D. Bettermann, S.F. Brady, T.H. Grossman, M.R. Liles, K.A. Loiacono, B.A. Lynch, I.A. MacNeil, M.S. Osburne, J. Clardy, J. Handelsman, and R.M. Goodman. 2000. Cloning the soil metagenome: A strategy for accessing the genetic and functional diversity of uncultured microorganisms. *Applied and Environmental Microbiology* 66:2541-2547.

Rontein, D., G. Basset, and A.D. Hanson. 2002. Metabolic engineering of osmoprotectant accumulation in plants. *Metabolic Engineering* 4:49-56.

Saliki, J.T., E.J. Cooper, and J.P. Gustavson. 2002. Emerging morbillivirus infections of marine mammals: Development of two diagnostic approaches. *Annals of the New York Academy of Sciences* 969:51-59.

Sanford, E. 1999. Regulation of keystone predation by small changes in ocean temperatures. *Science* 283:2095-2097.

Sanger, F., S. Nicklen, and A.R. Coulson. 1977. DNA sequencing with chain-terminating inhibitors. *Proceedings of the National Academy of Sciences USA* 74:5463-5467.

SCAR (Scientific Community on Antarctic Research). 2001. Report of the Subglacial Antarctic Lake Exploration Group of Specialists (SALEGOS). Meeting—I, Bologna, Italy, 29-30 November 2001. University of California, Santa Cruz.

Scherer, R. 2002. Sustained sea-ice free conditions in the Antarctic nearshore zone during marine isotope stage 31 (1.07 Ma). Abstract presented at American Geophysical Union's Annual Meeting, held December 5-10, 2002, San Francisco, California. Paper No. 73-14.

Schimel, J.P. 1995. Plant transport and methane production as controls on methane flux from Arctic wet meadow tundra. *Biogeochemistry* 28:183-200.

Schimel, J.P., and J.S. Clein. 1996. Microbial response to freeze-thaw cycles in tundra and taiga soils. *Soil Biology and Biochemistry* 28:1061-1066.

Schloss, P.D., L.R. Williamson, J.P. Kearns, J.F. Banfield, R.W. Ruess, W.D. Metcalf, and J. Handelsman. 2002. Alaskan soils: A cold microbial observatory. P. 68 in Moran, M.A., and S.L. Cadee (eds.), *Microbial Observatories/LexEn Principal Investigators' Workshop*. Arlington, Virginia: National Science Foundation.

Schoeberl, M.R., and D.L. Hartmann. 1991. The dynamics of the stratospheric polar vortex and its relation to springtime ozone depletions. *Science* 251:46-52.

Schreiber, U., W. Bilger, and C. Neubauer. 1994. Chlorophyll fluorescence as a non-intrusive indicator for rapid assessment of *in vivo* photosynthesis. In: Schulze ED, Caldwell MM (eds.), *Ecophysiology of Photosynthesis*. Berlin: Springer-Verlag.

Seaburg, K.G., B.C. Parker, R.A. Wharton, and G.M. Simmons, Jr. 1981. Temperature-growth responses of algal isolates from Antarctic oases. *Journal of Phycology* 17:353-360.

Serrano, R., J.M. Mulet, F. Rios, J.A. Marquez, I.F. de Larrinoa, M.P. Leube, I. Mendizabal, M. Pascual-Ahuir, M. Proft, R. Ros, and C. Montesinos. 1999. A glimpse of the mechanisms of ion homeostasis during salt stress. *Journal of Experimental Botany* 50:1023-1036.

Severinghaus, J.P., T. Sowers, E.J. Brook, R.B. Alley, and M.L. Benders. 1998. Timing of abrupt climate change at the end of the Younger Dryas interval from thermally fractionated gases in polar ice. *Nature* 391:141-146.

Shackleton, N.J., J. Backman, H. Zimmerman, D.V. Kent, M.A. Hall, D.G. Roberts, D. Schnitker, J.G. Baldauf, A. Desprairies, R. Homrighausen, P. Huddlestun, J.B. Keene, A.J. Kaltenback, K.A.O. Krumsiek, A.C. Morton, J.W. Murray, and J. Westberg-Smith. 1984. Oxygen isotope calibration of the onset of ice-rafting and history of glaciation in the North Atlantic region. *Nature* 307:620-623.

Shinozaki, K., and K. Yamaguchi-Shinozaki. 2000. Molecular responses to dehydration and low temperature: Differences and cross-talk between stress signaling pathways. *Current Opinion in Plant Biology* 3:217-223.

Sidell, B.D., M.E. Vayda, D.J. Small, T.J. Moylan, R.L. Londraville, M.L. Yuan, K.J. Rodnick, Z.A. Eppley, and L. Costello. 1997. Variable expression of myoglobin among the hemoglobinless Antarctic icefishes. *Proceedings of the National Academy of Sciences USA* 94:3420-3424.

Siegert, M.J., J.C. Ellis-Evans, M. Tranter, C. Mayer, J.-R. Petit, A. Salamatin, and J.C. Priscu. 2001. Physical, chemical and biological processes in Lake Vostok and other Antarctic subglacial lakes. *Nature* 414:603-609.

Siegert, M.J., M. Tranter, J.C. Ellis-Evans, J.C. Priscu, and W. Berry Lyons. In press. The hydrochemistry of Lake Vostok and the potential for life in Antarctic subglacial lakes. *Hydrological Processes*.

Small, J., S.R. Call, F.J. Brockman, T.M. Straub and D.P. Chandler. 2001. Direct detection of 16S rRNA in soil extracts by using oligonucleotide microarrays. *Applied and Environmental Microbiology* 67:4708-4716.

Smith, R.C., and K.S. Baker. 1979. Penetration of UV-band biologically effective dose rates in natural waters. *Journal of Photochemistry and Photobiology* 29:311-323.

Smith, W.O., Jr., and G.R. DiTullio. 1995. Relationship between dimethylsulfide and phytoplankton pigment concentrations in the Ross Sea, Antarctica. *Deep Sea Research* Part I 45:873-892.

Smith, R.C., B.B. Prézelin, K.S. Baker, R.R. Bidigare, N.P. Boucher, T. Coley, D. Karentz, S. MacIntyre, H.A. Matlick, D. Menzies, M. Ondrusek, Z. Wan, and K.J. Waters. 1992. Ozone depletion: Ultraviolet radiation and phytoplankton biology in Antarctic waters. *Science* 255:952-959.

Smith, R., L.C. Stapleford, and R.S. Ridings. 1994. The acclimated response of growth, photosynthesis, composition, and carbon balance to temperature in the psychrophilic ice diatom *Nitzschia seriata*. *Journal of Phycology* 30:8-16.

Smith, W.O., Jr., L.A. Codispoti, D.M. Nelson, T. Manley, E.J. Buskey, H.J. Niebauer, and G.F. Cota. 1991. Importance of *Phaeocystis* blooms in the high-latitude ocean carbon cycle. *Nature* 352:514-516.

Smith, W.O., Jr., J. Marra, M.R. Hiscock and R.T. Barber. 2000. The seasonal cycle of phytoplankton biomass and primary productivity in the Ross Sea, Antarctica. *Deep-Sea Research* 47:3119-3140.

Solomon, S. 1990. Progress towards a quantitative understanding of Antarctic ozone depletion. *Nature* 347:347-354.

Somero, G.N. In press. Thermal physiology and vertical zonation of intertidal animals: Optima, limits, and costs of living. *American Zoologist*.

Somero, G.N., and A.L. DeVries. 1967. Temperature tolerance of some Antarctic fish. *Science* 156:257-258.

Sowers, T. 2001. The $N_2O$ record spanning the penultimate deglaciation from the Vostok ice core. *Journal of Geophysical Research* 106:31, 903-914.

Spilliaert, R., and A. Gudmundsdóttir. 1999. Atlantic cod trypsin Y - Member of a novel trypsin group. *Marine Biotechnology* 1:598-607.

Stahl, D.A., and J.M. Tiedje. 2002. Microbial ecology and genomics: A crossroads of opportunity. American Academy of Microbiology, Washington, D.C.

Staley, J.T., R.L. Irgens, and R.P. Herwig. 1989. Gas vacuolate bacteria from the sea ice of Antarctica. *Applied and Environmental Microbiology* 55:1033-1036.

Stein, J.L., T.L. Marsh, K.E. Wu, H. Shizuya, and E.F. DeLong. 1996. Characterization of uncultivated prokaryotes: Isolation and analysis of a 40-kilobase-pair genome fragment from a planktonic marine Archaeon. *Journal of Bacteriology* 178:591-599.

Steponkus, P.L., M. Uemura, R.A. Joseph, S.J. Gilmour, and M.F. Thomashow. 1998. Mode of action of the COR15a gene on the freezing tolerance of *Arabidopsis thaliana*. *Proceedings of the National Academy of Science USA* 95:14,570-14,575.

Stockinger, E.J., S.J. Gilmour, and M.F. Thomashow. 1997. *Arabidopsis thaliana* CBF1 encodes an AP2 domain-containing transcriptional activator that binds to the C-repeat/DRE, a cis-acting DNA. *Proceedings of the National Academy of Sciences USA* 95:14570-14575.

Stolz, J.F., and R.S. Oremland. 1999. Bacterial respiration of arsenic and selenium. *FEMS Microbiology Reviews* 23:615-627.

Sunda, W., D.J. Kieber, R.P. Kiene, and S. Huntsman. 2002. An antioxidant function for DMSP and DMS in marine algae. *Nature* 418:317-320.

Suzuki, I., D.A. Los, Y. Kanesaki, K. Mikami, and N. Murata. 2000. The pathway for perception and transduction of low-temperature signals in Synechocystis. *EMBO Journal* 19:1327-1334.

Symer, D.E., C. Connelly, S.T. Szak, E.M. Caputo, G.J. Cost, G. Parmigiani, and J.D. Boeke. 2002. Human L1 retrotransposition is associated with genetic stability *in vivo*. *Cell* 110:327-338.

Tan, Z., T. Hurek, P. Vinuesa, P. Muller, J.K. Ladha, and B. Reinhold-Hurek. 2001. Specific detection of Bradyrhizobium and Rhizobium strains colonizing rice (Oryza sativa) roots by 16S-23S ribosomal DNA intergenic spacer-targeted PCR. *Applied and Environmental Microbiology* 67:3655-3664.

Tang, E.P.Y., R.Tremblay, and W.F. Vincent. 1997a. Cyanobacterial dominance of polar freshwater ecosystems: Are high-latitude mat-formers adapted to low temperature? *Journal of Applied Phycology* 33:171-181.

Tang, E.P.Y., W.F. Vincent, D. Proulx, P. Lessard, and J. de la Noüe. 1997b. Polar cyanobacteria versus green algae for tertiary waste-water treatment in cool climates. *Journal of Applied Phycology* 9:371-381.

Taylor, K.C., C.U. Hammer, R.B. Alley, H.B. Clausen, D. Dahl-Jensen, A.J. Gow, N.S. Gundestrup, J. Kipfstuhl, J.C. Moore, and E.D. Waddington. 1993. Electrical conductivity measurements from the GISP2 and GRIP Greenland ice cores. *Nature* 366:549-552.

Thomas, D.N., and G.S. Dieckmann. 2002. Antarctic sea ice—a habitat for extremophiles. *Science* 295:641-644.

Thomas-Keptra, K.L., S. J. Clemett, D.A Bazylinski, J.L. Kirschvink, D.S. McKay, S.J. Wentworth, H. Valli, E.K. Gibson, Jr., and C.S. Romanek. 2002. Magnetofossils from ancient Mars: A robust biosignature in the martian meteorite ALH84001. *Applied and Environmental Microbiology* 68:3663-3672.

Thomashow, M.F. 2001. So what's new in the field of plant cold acclimation? Lots! *Plant Physiolology* 125:89-93.

Torsvik V., and L. Ovreas. 2002. Microbial diversity and function in soil: From genes to ecosystems. *Current Opinion in Microbiology* 5:240-245.

Treonis, A.M., D.H. Wall, and R.A. Virginia. 2000. The use of anhydrobiosis by soil nematodes in the Antarctic Dry Valleys. *Functional Ecology* 14:460-467.

Turner, J., J.C. King, T.A. Lachlan-Cope, and P.D. Jones. 2001. Recent temperature trends in the Antarctic. *Nature* 418:291-292.

Turner, J. J.C. King, T.A. Lachlan-Cope, and P.D. Jones. 2002. Climate change (Communication arising): Recent temperature trends in the Antarctic. *Nature* 418:291-292.

# REFERENCES

Turtle, E.P., and E. Pierazzo. 2001. Thickness of a Europan ice shell from impact crater simulations. *Science* 294:1326-1328.
Urbach, E., K.L. Vergin, and S.J. Giovannoni. 1999. Immunochemical detection and isolation of DNA from metabolically active bacteria. *Applied and Environmental Microbiology* 65:1207-1213.
Vaulot, D., J.-L. Birrien, D. Marie, R. Casotti, M.J.W. Veldhuis, G. Kraay, and M.-J. Chretiennot-Dinet. 1994. Morphology, ploidy, pigment composition and genome size of cultured strains of *Phaeocystis* (Prymnesiophyceae). *Journal of Phycology* 30:1022-1035.
van Tienderen, P.H., A.A. de Haan, C.G. van der Linden, and B. Vosman. In press. Biodiversity assessment using markers for ecologically important traits. *Trends in Ecology and Evolution.*
Venter, J.C., M.D. Adams, E.W. Meyers, P.W. Li, R.J. Mural, G.G. Sutton, et al. 2001. The sequence of the human genome. *Science* 291:1304-1351.
Vernet, M. 2001. Effects of UV radiation on the physiology and ecology of marine phytoplankton. Pp. 237-278 in S. de Mora, S. Demers and M. Vernet (eds.), *The Effects of UV Radiation in the Marine Environment.* Cambridge University Press, Cambridge.
Vézina, S., and W.F. Vincent. 1997. Arctic cyanobacteria and limnological properties of their environment: Bylot Island, Northwest Territories, Canada (73°N, 80°W). *Polar Biology* 17:523-534.
Vincent, W.F., and M.R. James. 1996. Biodiversity in extreme aquatic environments: lakes, ponds and streams of the Ross Sea sector, Antarctica. *Biodiversity Conservation* 5:1451-1472.
Vincent, W.F., and P.J. Neale. 2000. Mechanisms of UV damage to aquatic organisms, Pp. 149-176 in S.J. de Mora, S. Demers, and M. Vernet (eds.), *The Effects of UV Radiation on Marine Ecosystems.* Environmental Chemistry Series. Cambridge University Press, UK.
Vincent, W.F., R.W. Castenholz, M.T. Downes, and C. Howard-Williams. 1993. Antarctic cyanobacteria: Light, nutrients, and photosynthesis in the microbial mat environment. *Journal of Phycology* 29:745-755.
Vinnikov, K.Y., A. Robock, R.J. Stouffer, J.E. Walsh, C.L. Parkinson, D.J. Cavalieri, J.F.B. Mitchell, D. Garrett, and V.F. Zakharov. 1999. Global warming and Northern Hemisphere sea ice extent. *Science* 286:1934-1937.
Vishnivetskaya, T., S. Kathariou, J. McGrath, D. Gilichinsky, and J.M. Teidge. 2000. Low-temperature recovery strategies for the isolation of bacteria from ancient permafrost sediments. *Extremophiles* 4:165-173.
Visser, I.K., M.F. van Bressem, M.W. van de Bildt, J. Groen, C. Orvell, J.A. Raga, and A.D. Osterhaus. 1993. Prevalence of morbilliviruses among pinniped and cetacean species. *Reviews in Science and Technology* 12:197-202.
Vorobyova, E., C. Soina, M. Gorlenko, N. Minkovskaya, N. Aalinova, D. Gilichinsky, E. Rivkina, and T. Vishnivetskaya. 1997. The deep cold biosphere: Facts and hypothesis. *FEMS Microbiology Review* 20:277-291.
Walsby, A.E. 1994. Gas vesicles. *Microbiological Reviews* 58:94-144.
Walter, M.A., D.J. Spillett, P. Thomas, and P.N. Goodfellow. 1994. A method for constructing radiation hybrid maps of whole genomes. *Nature Genetics* 7:22-28.
Wayne, R.K, J.A. Leonard, and A. Cooper. 1999. Full of sound and fury: The recent history of ancient DNA. *Annual Review in Ecology and Systematics* 30:457-477.
Weber, M.H.W., and M.A. Marahiel. 2002. Coping with the cold: The cold shock response in the Gram-positive soil bacterium Bacillus subtilis. *Philosophical Transactions of the Royal Society London B* 357:895-907.
Weckwerth, W., and O. Fiehn. 2002. Can we discover novel pathways using metabolomic analysis? *Current Opinion in Biotechnology* 13:156-160.

Weiler, C.S., and P.A. Penhale (eds.) 1994. *Ultraviolet radiation in Antarctica: Measurements and Biological Effects*. Antarctic Research Series Vol. 62. American Geophysical Union, Washington, DC.

Welsh, D.W., Y. Ishida, and K. Nagasawa. 1998. Thermal limits and ocean migrations of sockeye salmon (*Oncorhynchus nerka*): Long-term consequences of global warming. *Canadian Journal of Fisheries and Aquatic Sciences* 55:937-948.

Wharton, D.A., and M.R. Worland. 1998. Ice nucleation activity in the freezing-tolerant Antarctic nematode, *Panagrolaimus davidi*. *Cryobiology* 36:279-286.

Wharton, R.A., Jr., R.A. Jamison, M. Crosby, C.P. McKay, and J.W. Rice, Jr. 1995. Paleolakes on Mars. *Journal of Paleolimnology* 13:267-283.

Wheeler, P.A. 1997. Preface: The 1994 Arctic Ocean section. *Deep-Sea Research Part II-Topical Studies In Oceanography* 44:1483-1485.

Whyte, L.G., C.W. Greer, and W.E. Inniss. 1996. Assessment of the biodegradation potential of psychrotrophic microorganisms. *Canadian Journal of Microbiology* 42:99-106.

Willerslev, E., A.J. Hansen, and H.N. Poinar. In press. Experimental considerations on the recovery of DNA/RNA and viable organisms from ancient ice and permafrost. In J. Castello and S. Rogers (eds.), *Life in Ancient Ice*. Princeton Press.

Wu, L., D.K. Thompson, G. Li, R.A. Hurt, J.M. Tiedje, and J. Zhou. 2001. Development and evaluation of functional gene arrays for detection of selected genes in the environment. *Applied and Environmental Microbiology* 67:5780-5790.

Yamanaka, K. 1999. Cold shock response in *Escherichia coli*. *Journal of Molecular Microbiology and Biotechnology* 1:193-202.

Yang, H., E. Golenberg, and J. Shoshani. 1996. Phylogenetic resolution within the Elephantidae using fossil DNA sequence from the American mastodon (*Mammut americanum*) as an outgroup. *Proceedings of the National Academy of Sciences USA* 93:1190-1194.

Yayanos, A.A. 1986. Evolutional and ecological implications of the properties of deep-sea baraphilic bacteria. *Proceedings of the National Academy of Sciences USA* 83:9542-9546.

Yin, B., D. Crowley, G. Sparovek, W.J. de Melo, and J. Borneman. 2000. Bacterial functional redundancy along a soil reclamation gradient. *Applied and Environmental Microbiology* 66:4361-4365.

Yu, Z., G.R. Stewart, and W.W. Mohn. 2000. Apparent contradiction: Psychrotolerant bacteria from hydrocarbon-contaminated Arctic tundra soils that degrade diterpenoids synthesized by trees. *Applied and Environmental Microbiology* 66:5148-5154.

Zachos, J., M. Pagani, L. Sloan, and E. Thomas. 2001. Trends, rhythms, and aberrations in global climate 65 Ma to present. *Science* 292:686-693.

Zecchinon, L., P. Claverie, T. Collins, S. D'Amico, D. Delille, G. Feller, D. Georlette, E. Gratia, A. Hoyoux, M. A. Meuwis, G. Sonan, and C. Gerday. 2001. Did psychrophilic enzymes really win the challenge? *Extremophiles* 5:313-321.

Zehr, J.P., M.T. Mellon, and S. Zani. 1998. New nitrogen-fixing microorganisms detected in oligotrophic oceans by amplification of nitrogenase (nifH) genes. *Applied and Environmental Microbiology* 64:3444-3450.

Zehr, J.P., J.B. Waterbury, P.J. Turner, J.P. Montoya, E. Omoregie, G.F. Steward, A. Hansen, and D.M. Karl. 2001. Unicellular cyanobacteria fix N-2 in the subtropical North Pacific Ocean. *Nature* 412:635-638.

Zhao, Y., M. Ratnayake-Lecamwasam, S.K. Parker, E. Cocca, L. Camardella, G. di Prisco, and H.W. Detrich III. 1998. The major adult α-globin gene of Antarctic teleosts and its remnants in the hemoglobinless icefishes: Calibration of the mutational clock for nuclear genes. *Journal of Biological Chemistry* 273:14745-14752.

Zhou, J.Z., M.E. Davey, J.B. Figueras, E. Rivkina, D. Gilichinsky, and J.M. Tiedje. 1997. Phylogenetic diversity of a bacterial community determined from Siberian tundra soil DNA. *Microbiology UK* 143:3913-3919.

Zhu, J.-K. 2002. Salt and drought stress signal transduction in plants. *Annual Review of Plant Biology* 53:247-273.

APPENDIX A

# Committee Member Biosketches

**Dr. H. William Detrich III**, professor of biochemistry and marine biology at Northeastern University, earned his Ph.D. in biology from Yale University in 1979. His research focuses on the molecular adaptations that enable Antarctic fishes to survive and thrive in their cold, ice-laden marine environment. Specific areas of interest include the adaptation of the microtubule cytoskeleton and its associated motors to cold temperatures and the evolution of the Antarctic icefishes, the only vertebrate taxon that fails to make the oxygen transporter hemoglobin. Dr. Detrich has conducted more than 15 research expeditions to U.S. Antarctic research bases, including Palmer and McMurdo Stations, over the past 20 years. He currently serves on the Palmer Area Users' Committee and the Antarctic Research Vessel Oversight Committee, on groups that provide scientific advice to the U.S. Antarctic Program support contractor, and on the Group of Experts on Antarctic Biology of the Consiglio Nazionale delle Ricerche, which advises the Italian government. Working with the Polar Research Board, Dr. Detrich helped draft the proposal to protect two Sites of Special Scientific Interest near the Antarctic Peninsula: Western Bransfield Strait (SSSI-35) and Eastern Dallman Bay (SSSI-36).

**Dr. Jody W. Deming** is a professor of biological oceanography at the University of Washington. Dr. Deming earned her Ph.D. in microbiology from the University of Maryland in 1981. Her research interests include bacterial foraging and survival strategies, especially the use of extracellular enzymes in polar, sedimentary, and deep-sea environments;

assessing degradation of natural materials and organic contaminants in marine environments; and limits of microbial life in sea ice and the subsurface marine biosphere. Her interests include the molecular enzymatic basis for psychrophily in marine bacteria and its relevance to polar ecology, biotechnology, and bioremediation.

**Dr. Claire Fraser** is president and director of The Institute for Genomic Research (TIGR) and former director of the Department of Microbial Genomics and vice-president for research at TIGR. She earned her Ph.D. from State University of New York at Buffalo. As leader of the teams that sequenced the genomes of several microbial organisms, Fraser has helped initiate the era of comparative genomics. Her research interests include whole genome sequence analysis of microbial genomes and the use of genomic-based approaches to elucidate differences in gene expression. She currently serves on the Steering Committee for Exploring Horizons for Domestic Animal Genomics and the Science and Technology for Countering Terrorism: Biological Panel.

**Dr. James "Tim" Hollibaugh** is a professor at the University of Georgia and acting Director of the Department of Marine Sciences. He received his Ph.D. in oceanography in 1977 from Dalhousie University (Canada). His research interests include the structure and function of microbial communities, role of bacteria in biogeochemical processes, net ecosystem metabolism, polar oceanography, estuaries, and human impacts in the coastal zone. Dr. Hollibaugh also participated as a panelist in the National Research Council's workshop on Marine Biodiversity.

**Dr. William Mohn** is an associate professor of microbiology at the University of British Columbia. He earned his Ph.D. in microbiology in 1990 from Michigan State University. His laboratory is conducting studies aimed at developing technologies to biologically remediate Canadian Arctic sites contaminated with polychlorinated biphenyls (PCBs) and hydrocarbon fuels. He is part of a collaborative genomic project investigating *Rhodococcus* sp. RHA1. His lab also uses molecular approaches to elucidate and monitor complex microbial communities, including those in pulp mill wastewater treatment systems, polluted Arctic soils, and forest soils.

**Dr. John C. Priscu** is a professor of ecology at Montana State University, Bozeman. He earned his Ph.D. in microbial ecology in 1982 from the University of California at Davis and has worked on Antarctic systems for the past 18 years. His research focuses on biochemical transformations in polar freshwater and marine systems, physiological responses of microbes to icy environments, and the role of polar systems in global change research and astrobiology. He is currently the U.S. biology representative to SCAR, the Scientific Committee on Antarctic Research, and serves on the NSF Office Advisory Committee for the Office of Polar Programs and the NSF Advisory Committee for Environmental Research and Education. Dr. Priscu chairs a SCAR international group of specialists to outline plans for Antarctic subglacial lake research, including Lake Vostok and is a member of the U.S. ice core working group.

**Dr. George N. Somero** is the David and Lucile Packard Professor of Marine Science and the director of the Hopkins Marine Station of Stanford University. Dr. Somero earned his Ph.D. in biological sciences in 1967 from Stanford University, conducting research on Antarctic fishes. His research centers on the physiological, biochemical, and molecular mechanisms used by organisms to adapt to environmental variation, notably in temperature and ambient salinity. Current studies focus on amino acid substitutions that are important in adaptation of proteins to temperature, physiological determinants of biogeographic patterning, the physiology of invasive species, and the effects of environmental variation on gene expression. He previously served as a member of the Ocean Studies Board's Committee on Marine Reserves and Protected Areas and is a member of the National Academy of Sciences.

**Dr. Michael F. Thomashow** is a professor in the MSU-DOE Plant Research Laboratory and Departments of Crop and Soil Sciences and Microbiology and Molecular Genetics at Michigan State University. He earned his Ph.D. in microbiology in 1978 from the University of California at Los Angeles. His recent research has focused on the genetics of cold acclimation in *Arabidopsis* and other plants. He has discovered a family of regulatory proteins that control a battery of genes that impart both freezing and dehydration tolerance. These genes may have applications in increasing the freezing and drought tolerance of agronomic plants. He is a member of the American Academy of Microbiology and received the 2001 Alexander von Humboldt Foundation Award presented to the individual judged to have made the most significant contribution to American agriculture during the previous five years.

**Dr. Diana Wall** is a professor and Director, Natural Resource Ecology Laboratory, Colorado State University. She earned her Ph.D. in plant pathology in 1970 from the University of Kentucky. Her research focuses on assessing global change impacts on soil biodiversity and ecology in Antarctic Dry Valleys, the relationship of soil biodiversity to ecosystem functioning, nematode community structure and function, and consequences of human activities on soil sustainability. She chairs the U.S. Scientific Committee on Problems of the Environment (SCOPE), and serves on the U.S. National Committee for DIVERSITAS, the Committee on Agricultural Biotechnology, Health, and the Environment, and the U.S. National Committee for Soil Science.

APPENDIX B

# Workshop on Frontiers in Polar Biology Agenda

**Date:** September 9-10, 2002
**Location:** Granlibakken Conference Center and Resort, Tahoe City, CA

**September 9**
| | |
|---|---|
| 7:30 – 8:30 a.m. | Continental Breakfast |
| 8:30 – 8:40 | Welcoming Remarks |
| | H. William Detrich, III, Northeastern University |
| 8:40 – 9:10 | Plenary Talk I |
| | Donal Manahan, University of Southern California |
| 9:10 – 9:30 | Plenary Talk II |
| | H. William Detrich, III, Northeastern University |
| 9:30 – 10:00 | Task to the Breakout Groups |
| | H. William Detrich, III, Northeastern University |
| 10:00 – 10:30 | Break |
| 10:30 – 12:00 p.m. | Breakout Session A – "Research Priority and Application of New Technologies" |
| | • Identify high priority research questions that can benefit from new biological tools |
| | • Discuss applications of genomic science and functional genomics to polar bio-disciplines |
| | • Determine need for development of polar-specific technologies |
| 12:00 – 1:00 p.m. | Lunch |
| 1:00 – 2:30 | Breakout Session A (cont'd) |

| | |
|---|---|
| 2:30 – 3:00 | Break |
| 3:00 – 5:30 | Plenary A |
| | • Breakout groups report their findings |
| 5:30 – 6:00 | Plenary A Wrap-up |
| 6:00 – 7:00 p.m. | Reception |
| 7:00 – 8:30 p.m. | Dinner |

**September 10**

| | |
|---|---|
| 7:30 – 8:30 a.m. | Continental Breakfast |
| 8:30 – 10:00 | Breakout Session B – "Facilitate Transfer of Technology and Interactions among scientists" |

- Recommend ways to facilitate and accelerate transfer of genomic tools to polar research
- Seek ways to facilitate interaction between polar biological scientists and the broader community of biologist
- Assess impediments to polar genomics research
  —facilities
  —infrastructure
  —biological sampling collections
  —personnel and educational needs

| | |
|---|---|
| 10:00 – 10:15 a.m. | Break |
| 10:15 – 11:15 a.m. | Breakout Session II (cont'd) |
| 11:30 – 12:30 p.m. | Plenary B |
| | • Breakout groups report their findings |
| 12:30 – 1:30 p.m. | Lunch |
| 1:30 – 2:00 p.m. | Plenary B (cont'd) |
| 2:00 – 2:30 p.m. | Concluding remarks |
| | H. William Detrich, III, Northeastern University |

APPENDIX C

# Workshop on Frontiers in Polar Biology Participants

Bert Boyer, University of Alaska, Fairbanks
Michael Castellini, University of Alaska, Fairbanks
C.H. Christina Cheng, University of Illinois at Urbana-Champaign
Thomas A. Day, Arizona State University
John G. Duman, Notre Dame University
Andrew Gracey, Hopkins Marine Station, Stanford University
P. Michael Hasegawa, Purdue University
Brian Lanoil, University of California, Riverside
Donal T. Manahan, University of Southern California
Marcella A. McClure, Montana State University
Tsutomu Miyake, Virginia Mason Research Center
John C. Moore, University of Northern Colorado
Alison E. Murray, Desert Research Institute
Patrick J. Neale, Smithsonian Environmental Research Center
Donald R. Ort, University of Illinois at Urbana-Champaign
John Paul, University of South Florida
Lindsey Rustad, USDA Forest Service
Joshua Schimel, University of California, Santa Barbara
Stephanie Shipp, Rice University
Bruce Sidell, University of Maine
Blaire Van Valkenburgh, University of California, Los Angeles
Carol M. Vleck, Iowa State University
Naomi Ward, The Institute for Genomic Research

# APPENDIX D

# List of Acronyms

| | |
|---|---|
| AFGPs | antifreeze glycoproteins |
| AOB | ammonia-oxidizing bacteria |
| AOSB | Arctic Ocean Sciences Board |
| AOV | autonomously operated vehicle |
| ARCSS | Arctic System Science program |
| | |
| BAC | bacterial artificial chromosome |
| bp | base pairs |
| | |
| CASES | Canadian Arctic Shelf Exchange Study |
| CBF | CRT/DRE binding factor |
| CFC | chloroflourocarbon |
| COR | cold-regulated |
| Csp. | cold shock proteins |
| CT | computed tomography |
| | |
| DGGE | denaturing gradient gel electrophoresis |
| DMS | dimethyl sulphide |
| DMSP | dimethylsulphoneiopropionate |
| DNA | deoxyribonucleic acid |
| | |
| EC | Eocene |
| ERD | early response to dehydration |
| EST | expressed sequence tag |

APPENDIX D 165

| | |
|---|---|
| FIBR | Frontiers in Integrated Biological Research |
| FRF | fast-repetition-rate fluorometry |
| | |
| GC | guanine/cytosine |
| GC/MS | gas chromatography/mass spectrometry |
| GEN-EN | GENome-ENabled environmental sciences and engineering |
| GIS | geographical information system |
| GRC | Gordon Research Conference |
| | |
| HTC | high throughput culturing |
| HVRI | hypervariable region I |
| | |
| IAPP | International Arctic Polyna Programme |
| | |
| LDH | lactate deydrogenase |
| LEA | late embryo abundant |
| LexEn | NSF's Life in Extreme Environments Program |
| LINE | long interspersed nuclear element |
| LTER | Long-Term Ecological Research Program |
| | |
| MAAs | microsporine-like amino acids |
| MV | morbillivirus |
| | |
| NASA | National Aeronautics and Space Administration |
| NCEAS | National Center for Ecological Analysis and Synthesis |
| NICL | U.S. National Ice Core Laboratory |
| NRC | National Research Council |
| NSB | National Science Board |
| NSF | National Science Foundation |
| | |
| OPP | Office of Polar Programs (NSF) |
| | |
| PAC | P1 artificial chromosome |
| PAM | pulse amplitude modulated |
| PCR | polymerase chain reaction |
| PCSP | Polar Continental Shelf Project |
| PDP | pump during probe |
| PLFA | phospholipid fatty acid |
| RNA | ribonucleic acid |
| | |
| ROV | remotely operated vehicle |
| RTPCR | reverse transcription polymerase chain reaction |

| | |
|---|---|
| SCICEX | science ice expeditions |
| SHEBA | US-Canada Surface Heat Budget of the Arctic Ocean Program |
| SINE | short interspersed nuclear elements |
| SIP | stable isotope probing |
| SNP | single nucleotide polymorphisms |
| SOS | salt overly sensitive |
| ssu rRNA | small subunit ribosomal RNA |
| STS | sequence-tagged site |
| TEA | Teachers Experiencing Antarctica and the Arctic Program |
| UNOLS | University National Oceanographic Laboratory System |
| UV | ultraviolet |
| YAC | yeast artificial chromosome |